Advances in Actuators

Sensors Series

Series Editor: **B E Jones**

Other books in the series

Solid State Gas Sensors
Edited by P T Moseley and B C Tofield

Techniques and Mechanisms in Gas Sensing
Edited by P T Moseley, J O W Norris and D E Williams

Hall Effect Devices
R S Popović

Thin Film Resistive Sensors
Edited by P Ciureanu and S Middelhoek

Biosensors: Microelectrochemical Devices
M Lambrechts and W Sansen

Sensors VI: Technology, Systems and Applications
Edited by K T V Grattan and A T Augousti

Intelligent Sensor Systems
J E Brignell and N M White

Automotive Sensors
M H Westbrook and J D Turner

Thermal Sensors
Edited by G C M Meijer and S A W Van Herwaarden

Advances in Actuators

Edited by

A P Dorey
Lancaster University, UK

and

J H Moore
Lucas Bryce, Gloucester, UK

Institute of Physics Publishing
Bristol and Philadelphia

British Library Cataloguing-in-Publication Data

A catalogue record for this book is available from the British Library

ISBN: 0 7503 0291 7

Library of Congress Cataloging-in-Publication Data

Advances in actuators / edited by A. P. Dorey and J. H. Moore.
p. cm.—(Sensors series)
Includes bibliographical references and index.
ISBN 0-7503-0291-7
1. Actuators. I. Dorey, A. P. II. Moore, J. H. (John H.)
III. Series.
TJ223.A25A33 1995
629.8'315—dc20 94-47286
CIP

Series Editor: **Professor B E Jones**, Brunel University

Published by Institute of Physics Publishing, wholly owned by The Institute of Physics, London

Institute of Physics Publishing, Techno House, Redcliffe Way, Bristol BS1 6NX, UK

US Editorial Office: Institute of Physics Publishing, The Public Ledger Building, Suite 1035, Independence Square, Philadelphia, PA 19106, USA

Typeset by MCS Ltd, Salisbury, Wiltshire
Printed in the UK by J W Arrowsmith Ltd

Contents

Preface

The seeds of this work were planted in the mid-80s in the form of an internal company seminar held at the Lucas Group Research Centre in Shirley, West Midlands. The event was one of a series of 'technology updates' organized by the Research Centre and was aimed at a broad spectrum of engineers from all parts of the Lucas Group. Papers were presented by invited speakers drawn from many parts of industry and academia. The seminar was arranged by an old colleague, Chris Cullingford, who should take the credit for setting this exercise in motion.

Shortly after this meeting I found myself, as a member of the IOP Instrument Science and Technology (ISAT) Group, endeavouring to organize a similar event at IOP Headquarters in Belgrave Square. Naturally I was able to draw heavily on the Lucas seminar for my invited speakers. The event was well received and the proceedings subsequently published in the IOP Short Meetings Series as 'Actuator Update'.

Some time later Institute of Physics Publishing suggested to me that the whole area of actuator technology had been neglected in the technical press. After a period of cajoling, I was persuaded to undertake the task of expanding the original modest publication into a more balanced review of recent developments in actuator technology.

Although completely new to the problems of soliciting and editing a 'proper' text, I suspected that there would be a great deal more effort involved than was intimated and I was therefore greatly relieved to be joined in this task by my colleague and co-editor, Professor Tony Dorey.

In due course, the structure and contents of the text were agreed and a list of contributors identified. Some, to their credit, had been participants in the original Lucas seminar while others were more recent volunteers—for example, my own company links through Dr Kurt Petersen in Lucas NovaSensor CA enabled me to secure the distinguished contribution on 'Silicon Microactuators' from Dr Mehregany in Cleveland. My thanks to Kurt for his good offices in this respect.

My co-editor Tony Dorey was equally persistent in securing support from his contributors and the contents gradually began to take shape.

We both became well acquainted with the problems of deadlines and how to miss them (including our own!). Meanwhile life moved on and so did we.

I have moved my base of operations with Lucas from Cirencester to Birmingham and back to Gloucester, while Tony is Head of Engineering at Lancaster University. The lives of our contributors were equally complex and one or two fell by the wayside. Special thanks are therefore due to Dr Rod Greenough and his colleagues for stepping into the breach at the last minute with their contribution on 'Magnetostrictive Actuators'.

What was eventually needed was a strong hand to pull all of the loose ends together and IOPP rose admirably to the occasion. Thanks are due for their attention to detail, patience and perseverance when confronted with the excuses of contributors and editors alike!

In hindsight, I would like to think that the original observation was correct—the field of actuator technology will, I hope, benefit from this collection of expert contributions. A recent perusal of engineering journals does seem to confirm that actuators are now as newsworthy as sensors if not more so. Novel items appear all of the time—materials such as high-temperature superconductors are poised to make a significant impact over the next decade, the popular media are fascinated by the medical opportunities for microminiaturized machines, microactuation of large-scale mirror arrays looks certain to migrate from the world's most exotic optical telescopes into many more humble and widespread applications, and who knows what will arise from the field of nanotechnology?

The list goes on and on and I hope that here we have been able to bring together a snapshot of the current status of a few of the key developments.

My personal thanks are due to the contributors, the IOPP staff and my co-editor who have all had the enthusiasm and the patience to bring this text to fruition.

John Moore
Kingscote

Tony Dorey
Hornby

1

Introduction

A P Dorey and J H Moore

The development of machines has been largely associated with a human need to extend our physical capabilities to perform useful work—to convert energy in a convenient form into mechanical output in a controlled manner. The output of such machines, which is often physical motion (either linear or rotary) or a force of some form, extended the performance of the original prime movers, the human body or draught animals, to be able to move further or faster, or to be more powerful.

The Machine Age was therefore closely associated with the development of actuators in various forms, and a good deal of prosperity since the Industrial Revolution has been derived from various forms of actuator technology.

Early machine systems converted available energy, for example, thermal, gravitational or electrical energy, into a controlled mechanical output, and exercised some degree of control over the response of the system to variations in input parameters or loads. The overall control of system behaviour was relatively crude, with inputs supplied by simple sensing means and the control function itself often also residing in mechanical arrangements—or the hand and eye skills of a human operator.

However, it would not be true to imply that machine systems remained crude, and as their development progressed then the complexity and sophistication of the sensing and control functions within the system were also extended, often resulting in elegant and ingenious integration of system inputs and control elements within the actuator itself. None the less, the key element remained the actuator itself and the fundamental objective was to convert energy into mechanical work.

More recently, computer systems have provided the impetus for the so-called 'Second Industrial Revolution', in which information as much as

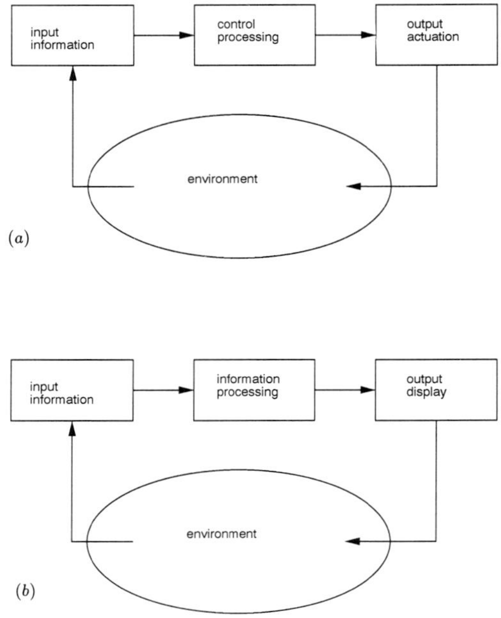

Figure 1. (*a*) Basic machine system. (*b*) Basic information processing system.

energy has become a prime commodity to be processed. Facilitated by spectacular advances in microelectronic component technology from the early 1960s onwards, particularly in the development of microprocessors and solid state memory devices, there has been a major shift in emphasis. System developers took much more interest in extending the processes of information input and control. In the past 20 years substantial worldwide activity has focused on the technologies of machine input (sensors) and information processing (control) while the actuator field has lain relatively dormant, with perhaps the most notable exception being intensive developments of actuators associated with computer peripherals themselves.

Throughout this period, actuator development has remained something of a Cinderella of the new microelectronic age.

An indication of the relative scarcity of activity during this period can be gleaned from a perusal of the technical journals. For example, the journal *Sensors and Actuators* (Elsevier Sequoia SA) began publication in 1981. In the period 1981 to 1990 (volumes 1 to 24) only about 7% of papers published were concerned with the technology of actuators. Significantly, that situation is now showing every sign of change and actuator technologies feature much more prominently in the technical press. New developments are emerging from laboratories worldwide, and it is an aim of this text to provide a sample of some of these developments and an indication of the continuing opportunities for innovation in this field.

Some innovations arise from the exploitation of the characteristics of new materials, while others have emerged as a consequence of new and sophisticated materials processing techniques. Even in the most conventional technical areas there are real signs that innovation continues to bear fruit.

A process of actuator technology classification can give a useful insight into the underlying science. Inputs and outputs may be classified by energy inputs and outputs in a process analogous to that adopted by Middlehoek and Audet in *Silicon Sensors* chapter 1 (published by Academic Press). Generally, however, we are interested only in outputs in a mechanical form and these may be subdivided into motion (linear or rotary) and force (linear or torque). An alternative subdivision might be to distinguish between the primary forms of output media—gaseous (pneumatic), liquid (hydraulic) and solid state (e.g. piezoelectric).

Figure 2 shows generalised energy flows in an actuator and is the basis for this particular classification scheme. The primary input energy for the actuator will, in the main, be derived from electrical, hydraulic, pneumatic, thermal or mechanical sources.

The input or control information for an actuator may take a wide variety of energy forms, particularly as non-electrical means of information transmission gain in popularity.

Developing along these lines a two-dimensional classification matrix can be derived, as in figure 3. The exercise of completing this matrix might be regarded as academic, but is none the less thought provoking.

New actuator technologies may be derived from a number of broad fields of innovation.

(*a*) Characteristics of new materials extending known effects to new levels (e.g. Terfenol, a material exhibiting giant magnetostriction) or new forms of response (e.g. shape memory alloys).

(*b*) New materials processing techniques. Here, obvious examples are the considerable developments associated with silicon micromachining. These, originally associated with the development of silicon sensors, are

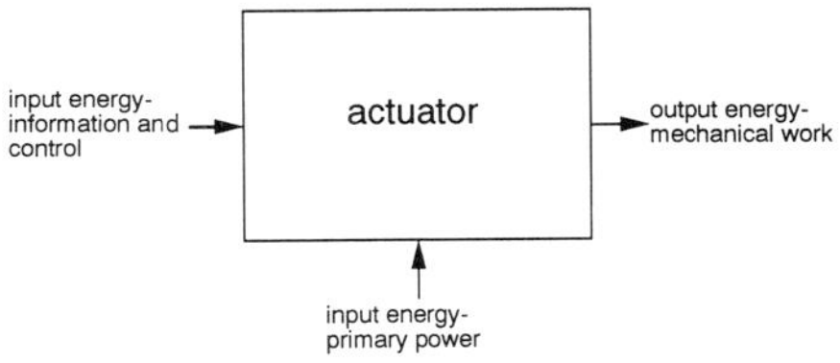

Figure 2. Basic energy flows.

control i/p \ power i/p	electrical	pneumatic	hydraulic	thermal	mechanical
electro-magnetic	electro-magnetic magneto-strictive piezoelect	servo-valves	servo-valves	shape memory alloys	electro-rheological fluids
pneumatic	electro-pneumatic devices	pneumatic amplifier + piston			
hydraulic	electro-hydraulic devices		hydraulic amplifier+ piston		
thermal	shape memory alloys			shape memory alloys bimetals	
mechanical	mechanical control of electro-magnetic machines	mechanical control of pneumatics	mechanical control of hydraulics		gearboxes
radiant	optical actuator	optical actuator	optical actuator		

output = motion or force

Figure 3. Classification of actuator technologies.

now opening the way for the development of microactuators and indeed complete micromachine systems.

(*c*) New media for the processing and transmission of information, leading to new forms of actuator input, e.g. optical interfaces with actuators.

(*d*) Ingenious developments in an established field. There is always room for the new idea no matter how commonplace the technology. A good example here is the recent development of simple pneumatic actuators based on restrained bladders emulating the behaviour of human muscles, which will find ready application in automated models of the human figure.

In due course, some actuators will develop into systems in their own right as the reduction of the size and costs of electronic functions encourages the extensive use of 'distributed' intelligence—the ubiquitous 'smart' component. It is reasonable to anticipate that digital communications, fault monitoring and diagnosis and self-calibration will all have a role to play. It will become increasingly difficult to distinguish between 'components' and 'systems'.

This text is intended to provide a worthwhile and stimulating 'snapshot' of some aspects of actuator developments in the early 1990s. Some technologies are more mature than others. In some cases spheres of application are obvious while others appear to be more like 'solutions looking for problems'. But then, so was the laser in its early years. Each chapter has been contributed by acknowledged experts in their field and is self-contained, however the chapters fall into two broad groups. The first group (chapters 2–4) is concerned with new technologies applied to generally 'conventional' areas of actuator application while the second (chapters 5–8) deals with perhaps more novel fields of activity.

Chapter 2, *Fluid Power Actuators*, opens the text by reviewing what might be expected to be one of the most well developed areas of actuator technology. Popular interest has recently been aroused not so much by basic technologies as by the latest areas of application. In particular, both pneumatic and hydraulic actuators have found their ways into various forms of motor vehicle suspension in combination with sophisticated electronic control systems. One of the earliest and still one of the most significant developments in this area was the fully active hydraulic suspension system produced by Lotus Engineering. Initially developed for Grand Prix cars, the system is confidently expected to make an impact on the consumer market in due course. Pneumatic suspension systems have also made an appearance on vehicles and the latest Range Rover off/on highway vehicle features an electronically controlled pneumatic ride height system.

Remaining on the theme of hydraulic actuation, chapter 3, *Electro-rheological Actuators*, details the status and potential for the use of these novel materials in actuation applications. It is exciting to note that many of their early weaknesses such as environmental limitations are now being eliminated. There seems to be real potential for the development of actuators with very few moving parts, based on ER fluids.

In any real applications of actuators, it is impossible to ignore the role played by the electronic elements which frequently control the actuator. Indeed the integration of control element with actuator leads inevitably towards the 'Smart Actuator' concept. Chapter 4, *Smart Power Devices*, reflects the key role played by these control elements. This chapter also illustrates the environmental demands placed on electronic devices integrated into actuators. Whereas many electronic components are normally carefully protected from the worst excesses of thermal, mechanical and electrical environments, the interface role and locations of actuators frequently ensures that the opposite is true. 'Smart Actuators' may prove to be a substantial challenge for the designers of 'Smart Power' devices.

Clearly one of the major developments in recent years has been the move from optical signal transmission by fibre optics towards the possibility of complete signal processing in the optical domain. The advantages of the optical medium for achieving intrinsic safety and immunity to EMC are well known. It is not surprising that the potential for optical actuation is receiving attention and chapter 5, *Optical Actuators*, gives an excellent account of the status of this technology.

The techniques developed for the manufacture of microelectronic components have generated a vast range of processes concerned with the 'very small'. Lately this field of interest has moved even beyond the immediate demands of the microelectronics industry and into the new area of nanotechnology. Chapter 6, *Miniature Nanoactuators*, reviews many of the technologies available for the achievement of controlled motion in the nanometer range. It is interesting to realize that these devices are already opening up new approaches to the fundamental experimental study of the laws of quantum physics while at the same time providing new techniques to relate fundamental standards of length to crystallographic dimensions. Many new and exciting developments will emerge from this field.

Remaining with the microelectronic industry, chapter 7, *Silicon Microactuators*, concerns itself with many novel devices fabricated using the latest sophisticated silicon processing technologies. A whole range of miniaturized machines is emerging from the laboratory—motors, mechanisms, microvalves etc. Initially developed as a means of exploring the potential of the technology itself, they soon began to lead to insights into the mechanical behaviour of surfaces at the atomic level. Recently there have been reports of thermal microvalves capable of providing a fluidic equivalent to the transistor, and of course many people anticipate that miniaturized machines will eventually have a major role to play in advanced medical techniques.

Finally, chapter 8, *Magnetostrictive Actuators*, returns once more to the theme of enhanced materials characteristics leading to new generations of actuator. The development of a material with 'giant' magnetostrictive response is described, and an exciting range of potential applications is described—from sonar devices to the control of internal combustion engine exhaust valves.

Overall, the actuator field appears to be moving forward quite rapidly. New and novel technologies are now reported frequently in the pages of the technical press. For example recently *Eureka* has reported *Shape Memory Alloy Actuators* (December 1991 p 20), *Pneumatic Equivalents to Human Muscles* (July 1993 p 5), *Novel Combinations of Pneumatics and Linear Electromagnetic Machines for Long Stroke Actuators* (July 1993 p 23). Opportunities for innovation appears to be endless. For the future, the interested reader is left to speculate on new directions and products arising particularly from micromachines and nanoactuators. What will be their real potential? What new markets will emerge?

We started originally by noting actuators as a response to a human need for extension of muscle power and we are now confronted by reductions of size and action down towards the atomic scale. Who can fail to be intrigued by the capability of the scanning tunnelling microscope (an actuator-based invention of recent years) to both image and manipulate the individual atoms in the surface of a solid?

Whatever the technical possibilities however it will always be important to recognize that laboratory demonstrations alone do not automatically lead to commercial success. The needs of the market place must be identified and the requirements of the individual customers must always be satisfied. Novel technology alone is no guarantee of commercial success outside the academic world and today's product engineers must take full account of customer needs if the challenges of global competition are to be met.

2

Fluid Power Actuators

David Dawson

2.1 INTRODUCTION

Hydraulic and pneumatic systems are commonly regarded as comprising the full scope of fluid power systems, although there are more particular techniques, such as the use of ferrofluids and electrostatic fluid-based devices used in power transmission products such as clutches. Aspects of these are discussed in chapter 3. A distinction also needs to be made between fluid power systems in which the static pressure of the working fluid acts to provide the motive force in the actuator, and those in which a change of momentum in the working fluid produces a force or torque in a mechanical device.

The first category comprises the great majority of fluid power applications, the actuators being typically hydraulic or pneumatic rams and motors and their associated controls. This chapter deals with these systems. The second category includes devices such as torque convertors and pelton wheels and is exclusively rotational with a fairly coarse control of speed or torque sufficing for most purposes, and as such, is omitted from this chapter. Readers interested in these systems will find adequate descriptions in standard texts [1].

Any design engineer facing an actuation requirement is confronted by a plethora of possibilities not just in terms of detailed configuration or supplier, but quite often also in fundamental operating principle. Hydraulic and pneumatic systems frequently appear to compete directly with electrical systems and the best solution cannot be discerned without the expenditure of much thought on the detailed parts of the specification. The dilemma may remain unresolved even up to the final embodiment as witnessed by the entirely satisfactory but 'competing' systems, hydraulic cylinders or electrically driven ballscrews, for positioning aircraft wing flaps.

This chapter gives some guidance on how to approach the issue of actuator selection, together with some pointers to new developments in the control of actuators in fluid power systems.

2.2 MATCHING THE ACTUATOR TO THE SPECIFICATION

Consideration of a range of applications from, for example, aircraft wing flaps (a large mobile system) through a packaging conveyor (a plant-based system) to a contact lens lathe (a stand-alone precision machine), reveals a large number of quantifiable and qualitative parameters which may have to be taken into consideration. It is hardly possible to put these into an order of priority, but a broad classification may be attempted thus:

Functional

Stroke or angle
Velocity or rotational speed
Force or torque
Inertias—self- and load
Efficiency
Thermal effects—heat dissipation and cooling
Position measurement
Control
Acceleration or retardation profiles
Frequency response
Guidance
Hold load on power-off
Backlash
Stiffness
Repeatability
Resolution

Configuration standards, safety and maintenance

Mounting
Power source
Cost
Size
Applicable standards
Safety
Noise
Failure modes

Protection
Interference
Reliability
Environmental factors and materials
Maintenance

The initial matching will be through an examination of the primary quantifiable criteria, and in such an approach range charts, in effect well-digested synopses and concatenations of suppliers catalogues, will be of assistance. The apparent range of options may be of such extent as to result in the choice of idiosyncratic non-optimal solutions. Research into the process of design, building up schemes from proprietary components, is pointing toward the use of automated matching and simulation techniques to aid the engineer in the early stages of conceptual design [2].

The consideration of fluid power-based actuators will take into account their range of application and the above criteria, recognizing that in most

Table 1. Comparison of pneumatic and hydraulic systems.

Pneumatic systems	Hydraulic systems
Uses: Packaging machinery Safety guards Pick-and-place 'robots' Textile machinery	Uses: Presses Winches Machine tools Aircraft controls
Force: Up to 10^4 N	Force: Over 10^8 N
Higher velocity/rotational speed	Lower velocity/rotational speed
Stroke: Up to 5 m	Stroke: Generally under 2 m
Light-alloy construction	Heavy-steel construction
Polymer tubing (nylon) connections to actuator	Steel pipe or high-pressure hose connections to actuator
No return connection; exhaust to atmosphere	Requires return line and in some actuators, casing drain line too
Intrinsically compliant, can only hold load against a mechanical stop	Intrinsically stiff, precise control of position is possible with sensor and feedback loop
Low efficiency of energy utilization	Usually high efficiency of energy utilization
Comparatively cheap and easy to install	Much more expensive and requires a higher level of skill to install

cases the terms will only have meaning if the actuator itself is regarded as part of a 'mechatronic' system including sensing and control.

The special features of each type of fluid power actuator/system are dealt with in the detailed sections following, but a short summary of their general characteristics and application fields, given in table 1, may be useful at this point.

2.3 FLUID POWER SERVICES

All actuation systems involve the conversion of some form of energy into mechanical work. In considering practical systems, it is seen at once that there may be several stages of energy conversion and if efficiency is an important criterion in the selection of an actuator, it may be necessary to consider the whole system, for example right back to the point at which the electrical power cables come through the factory fence, including the losses at each stage of conversion and transmission. Evidently, different criteria will apply where we are dealing with mobile machinery units such as hydraulic excavators driven by diesel engines and packaging machines incorporating pneumatic cylinders supplied from a factory air main. Where a fluid power distribution system already exists, there will obviously be a predisposition to use it as the apportioned costs of the distribution system may be marginal.

Both hydraulic and pneumatic systems are used in environments where a fire hazard may exist, although the flammability of normal hydraulic mineral oils may present a problem. Synthetic oils such as phosphate esters are sometimes employed but at greatly increased cost, hence the current interest in water-based hydraulic fluids. Compressed air of course is ideal for use in actuators close to heat and flame in environments such as glassworks.

2.3.1 Compressed Air Systems

These are found widely in process plant and materials handling operations such as bottling plants. End applications of pneumatic actuators include rams, control valve positioners and (less commonly) as air motors in rotational drives. The primary reasons for using compressed air systems are:

Convenience—the working fluid itself is free, so there is no need for a return line.

Cleanliness—leakage is non-polluting.

Intrinsic safety—useful where there are fire and explosion risks.

For industrial purposes, most systems are based on a standardized working pressure of 7 bar (0.7 MPa) gauge which is set by the compressor control system. Different configurations of compressor are appropriate to a range of flow rates as given in table 2.

In most cases however, even with the addition of interstage cooling, the overall efficiency of compression may be as low as 30%. Further losses will occur in the air distribution system. Industrial air systems will always incorporate means for extracting oil and condensate from the air supply which would otherwise tend to clog instrument and control orifices, unless the compressors are of a type which is intrinsically oil-free. Adjacent to actuators however, it is usually necessary to provide an in-line lubrication system to ensure that mating parts do not run dry. For similar reasons, air filtration to provide removal of dust particles is always provided at the machine inlet and sometimes at the outlet as well.

Compressed air systems are also found in applications other than industrial plant or large mobile installations such as ships. For example, virtually all EEC heavy goods vehicle trailers now incorporate both air-controlled braking systems and air-spring suspensions. Increasingly responsive electropneumatic air control devices now allow pneumatic diaphragm actuators to be employed in safety-critical areas such as anti-lock braking (ABS). In these systems the same consideration must be given to the maintenance of cleanliness as in large industrial systems. In passenger vehicle suspensions the provision of an air compressor may be a serious on-cost and energy drain.

2.3.2 Hydraulic Systems

The main differentiation between hydraulic and pneumatic systems is in working pressure, with hydraulics typically running at between 100 and 200 bar (10 to 20 MPa). Also hydraulic systems tend to be supplied by unit power packs in which a pump supplies a small number of actuators, perhaps

Table 2. Application ranges of air compressors.

Machine type	Typical delivery ($m^3\,min^{-1}$)	Typical pressure bar ($10^5\,N\,m^{-2}$)
Piston	0.05–50	2–15
Vane	0.25–15	2–7
Water ring	2.00–25	2–7
Rotary screw	2.00–100	2–10
Centrifugal	50.00–500	1–10

only one. Distributed or ring-main systems are fairly uncommon apart from special cases such as in ships or aircraft.

Whereas air compressors are differentiated chiefly by ranges of volumetric flow rate, hydraulic pumps are differentiated by pressure according to the boundaries shown in table 3. The most common type of industrial hydraulic pump is the axial piston machine (figure 1), which in its common pressure-compensated variable displacement form is the basis of a power pack comprising the pump itself, drive motor, reservoir and filtration system. The condition of the working fluid (usually a light mineral oil) is critical in hydraulic systems because of the close working clearances employed in such items as electrohydraulic control valves.

Hydraulic systems share some of the positive features of pneumatic systems; they are favoured for use in hazardous environments and the use of pressure relief valves local to the actuator will protect both the load and the

Table 3. Application ranges of hydraulic pump/motors.

Type of machine	Typical delivery litres (s^{-1})	Typical pressure bar ($N\,m^{-2}$)
Gear	0.2	0.2–100
Vane	1.0	10–100
Radial piston	2.0	50–150
Axial piston	5.0	100–300

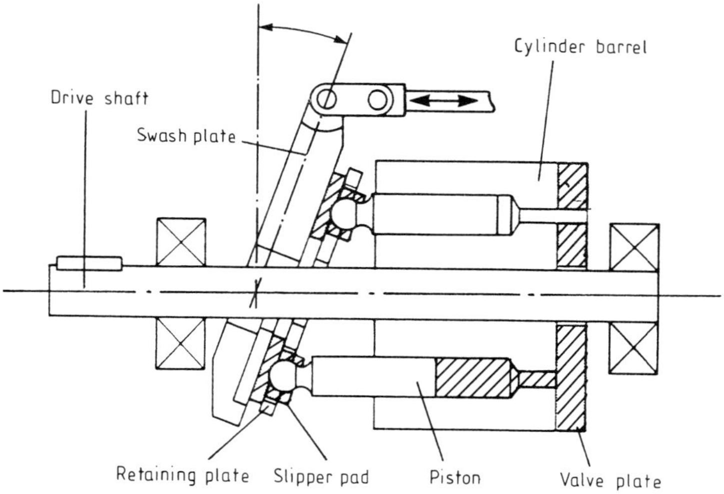

Figure 1. Variable displacement axial piston machine. (Reproduced by permission of Chapman and Hall.)

actuator itself against damage under overload or jamming conditions. The primary reason for using hydraulics however is the very high force achievable within a small bulk and the very flexible layout configurations which are possible.

Poor design and installation practice has too often given hydraulics systems, at least in general industry, an undeserved reputation for leakage, contamination and noise.

2.4 PNEUMATIC ACTUATORS

2.4.1 Linear Pneumatic Actuators

These are the commonest form of pneumatic actuator as typified by the pneumatic ram (figure 2), which is a double-acting unit with end cushioning to provide shock absorption at the end of the stroke. Such rams are commonly used in 'bang-bang' two-position configurations and are operated by a simple directional control valve. Rams may also be single acting in which case the return motion is provided by the opposing force in the system, by gravity or by an internal return spring.

The scope of application of pneumatic linear actuators has been greatly increased in recent years by the introduction of rodless air cylinders as shown in figure 3. These are available with strokes up to 5 m or even more, and are an alternative to the use of rotational drives with racks or chains. Whereas it is usually necessary with conventional rams to provide a separate means of guiding the load and preventing side forces from being applied to the ram, with these units, the guidance function is built in. A further advantage is that 80% or more of the overall or extended length of the device is available as working stroke. With conventional rams it is under 45%.

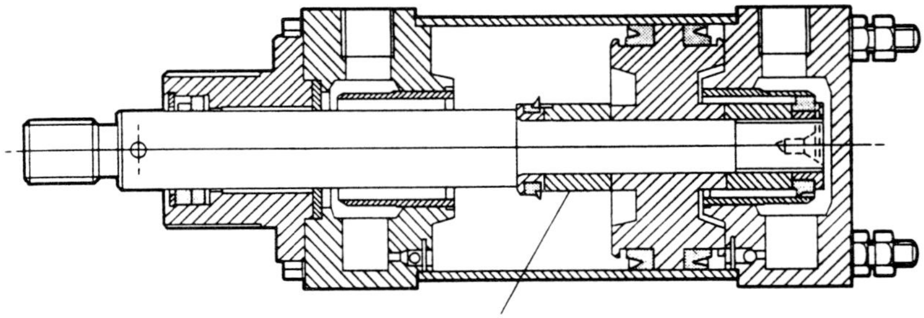

Figure 2. Cushioned pneumatic ram (Martonair). (Reproduced by permission of Chapman and Hall.)

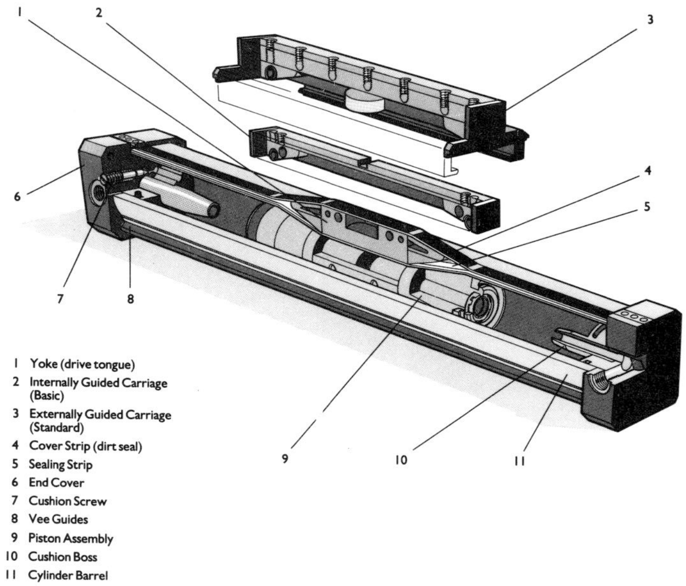

Figure 3. Rodless air cylinder (Martonair).

Large forces, 10^4 N or more, may be obtained using diaphragm-type actuators (figure 4), but with a short stroke of 50 mm or so. These are popular in the process industry for operating pipeline control valves, and incorporate special control units able to position the valve stem to acceptable levels of accuracy using pressure (and hence force) modulation against a large spring. Light duty and instrumentation applications are met by means of a bellows or aneroid-type capsule.

2.4.2 Rotational Pneumatic Actuators

The most common application is of *limited* rotation, commonly used for operating plug-type fluid control valves requiring nominally 90° of rotation from 'on' to 'off'. These requirements are met by vane-type actuators (figure 5), or opposed rack and pinion machines as in figure 6 which are capable of three or four complete rotations.

Continuous rotation devices, or air motors, are commonly employed in hand power tools such as drills, nut runners and torque wrenches where their ability to run to a stalled condition without damage is of great advantage. Speed control of pneumatic motors is problematic since it is highly load-dependent and can only be achieved by throttling. The output torque is of course directly proportional to the input pressure which can easily be controlled. This feature is also an advantage in specialized high-power

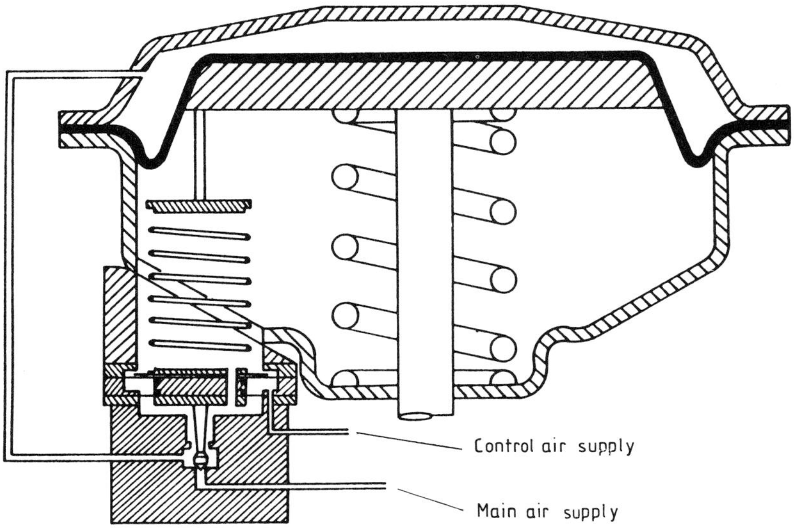

Figure 4. Air diaphragm actuator. (Reproduced by permission of Chapman and Hall.)

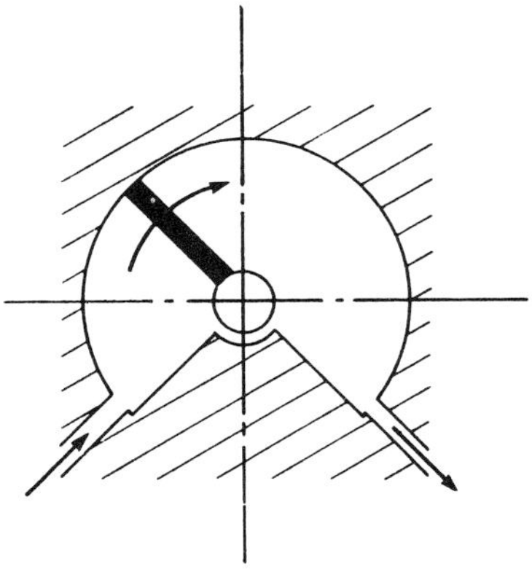

Figure 5. Vane actuator. (Reproduced by permission of Chapman and Hall.)

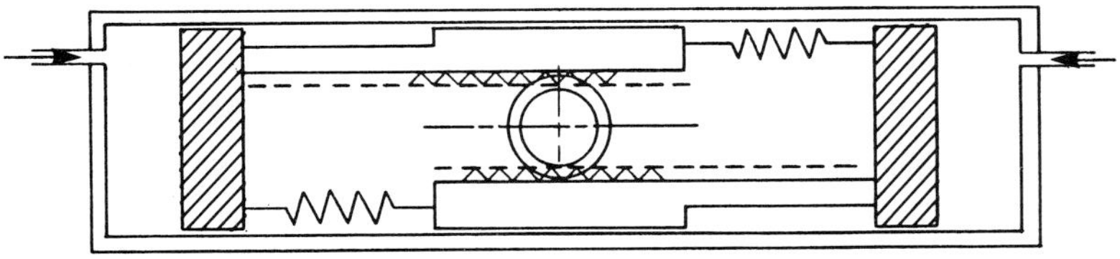

Figure 6. Balanced opposed piston actuator. (Reproduced by permission of Chapman and Hall.)

drives such as rock drills used in mines and quarries where the exhaust air can also be used to clear the bit. Otherwise it is rare in industry for air motors to be used for continuous operation in other than very low-power applications (up to 250 W) because of their intrinsically low efficiency. A typical example would be for laboratory stirrers.

2.4.3 The Control of Pneumatic Actuators

Pneumatic actuators are intrinsically 'springy' due to the compressibility of the working medium. This is used to advantage in vehicle suspensions, but is a severe drawback in the majority of applications. The control of rotational drives is achieved by the use of inlet throttling with a valve (figure 7), which is designed to control the pressure applied to the actuator. Such valves employ the principle of force balance with the controlled pressure acting on a diaphragm in such a way as to restrict the air supply. The set point is adjusted manually by means of a screw which varies the precompression in the control spring, or remotely by a stepper motor. Such methods of control are most satisfactory where the load has a strongly rising torque/speed characteristic (figure 8), but are to be regarded as effective rather than efficient. Large losses may be incurred in the valve.

Most industrial applications of pneumatic actuators are of cylinders and control is most usually elementary 'bang-bang' position control in which the cylinder is driven to end stops at either end of its travel. The system designer must normally ensure that the external forces from the load do not at any time exceed the force provided by the cylinder, unless there is a safety

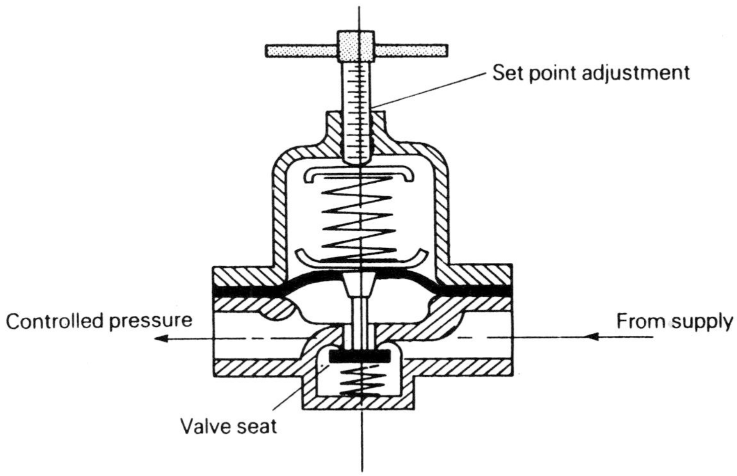

Figure 7. Pressure control valve. (Reproduced by permission of Chapman and Hall.)

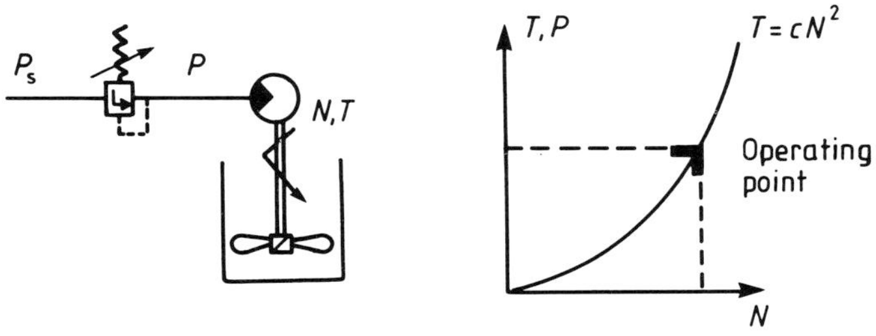

Figure 8. Throttling valve control of pneumatic motor.

requirement for allowing the cylinder force to be overcome. Spring-centred three-position solenoid-operated spool valves are commonly employed in the two-stage (piloted) form (figure 9). For convenience, these are often mounted on a common base plate for compactness and to allow the air supply to be manifolded. The sizing of valves is somewhat empirical and manufacturers' catalogues use a flow factor K_v which at a specified pressure difference will give the flow rate in $1\ s^{-1}$ referenced to standard air conditions. The effect of undersizing of valves and of the interconnecting piping is to introduce a possibly undesirable delay into the operation of the actuator.

It is possible to build up complex pneumatic circuits using fluidic elements for two-position control of actuators. These elements are based upon the Coanda, or wall attachment, effect in which a jet of fluid emerging from an orifice close to a surface will remain attached to the surface unless disturbed. In the example shown of a bistable (figure 10), the disturbance is provided by a pulse at the control port A, causing the jet to detach from the adjacent wall and changing the output from port 1 to port 2. Within fluidics, it is possible to emulate many of the components of electronic logic circuits with AND, NOR and NOT gates and cascade arrangements providing large signal amplification and switching suitable for driving air cylinders.

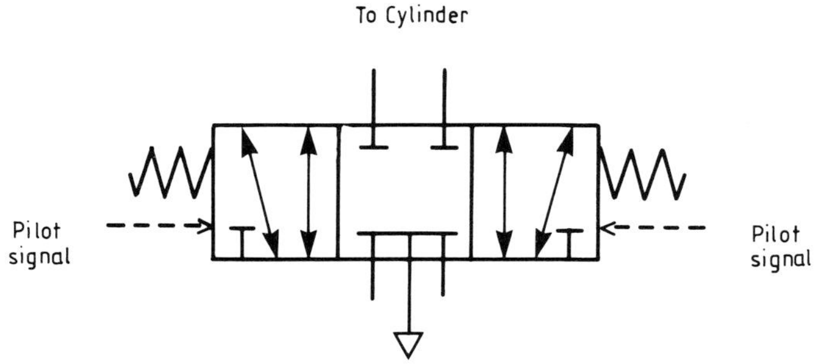

Figure 9. Second-stage piloted valve.

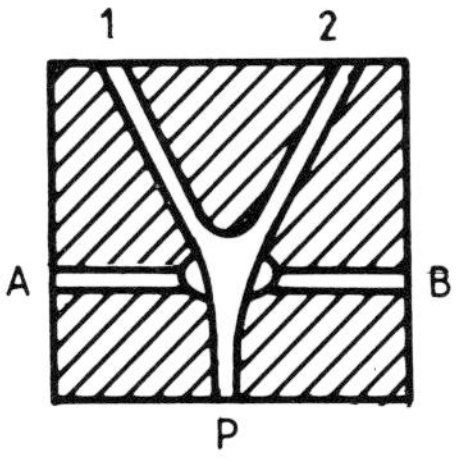

Figure 10. Fluidic bi-stable.

Manufacturers such as Crouzet and Telemecanique are among the principal suppliers of this type of equipment. Further reading may be found in [3]. Fluidic techniques were very popular in the 1970s in areas such as petroleum refineries where the electronic controls then available were viewed less favourably because of the danger of spark hazards. With the present use of integrated circuit-based electronic control systems of much lower energy density, that objection has been substantially reduced. Fluidic components are also very sensitive to contamination, therefore good standards of air quality are essential. An alternative to fluidics, different in principle but similar in effect, is 'moving-part logic'. Here the fluidic elements are replaced by miniature poppet or spool valves (figure 11), which are operated to switch outputs in the same manner as the fluidic element. These are said to be less susceptible to contamination, but good filtration is still essential.

2.4.4 Recent Advances in the Control of Pneumatic Actuators

Recent developments in pneumatic controls may be roughly classified into two areas: those which seek to improve the performance of controls for air cylinders; and less generic implementations which are specific to products, many of which are associated with vehicles.

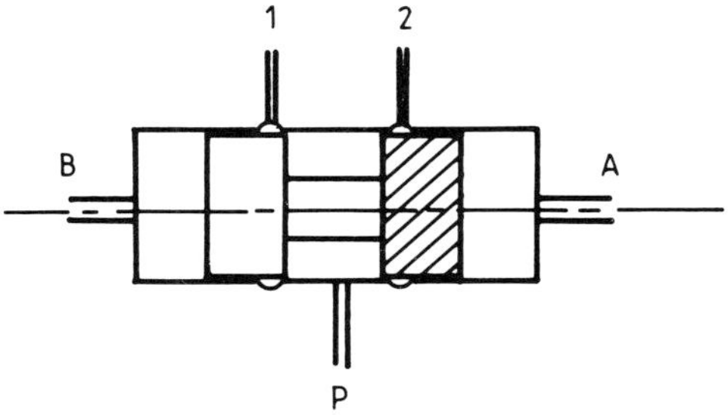

Figure 11. Moving part logic element.

Proportional position-controlled pneumatic cylinders As mentioned above, pneumatic cylinders have been used primarily in 'bang-bang' mode. An attractive objective has been to find a means of controlling a cylinder at any point within its working range. This has now become possible through the use of rapid-acting proportional pressure control valves (figure 12), which can be employed in a feedback control circuit. A single valve may be used if a spring return is employed or if the load is otherwise unidirectional. For double-acting cylinders, such valves may be used in pairs or alternatively a $\frac{4}{3}$ proportional valve may be used as in figure 13. Here, a disturbing external force displaces the cylinder and sensor which acts through the controller and valve to produce an increase of pressure, and oppose the disturbing force. There are obvious limitations on the responsiveness of such a system which should normally be employed in quasistatic applications, or at least where the disturbing frequencies are not more than a few Hz. Cylinders are however available with cam action rod brakes which allow a position, once

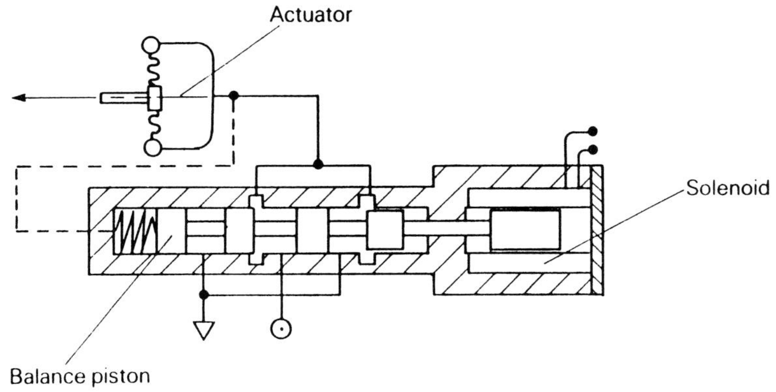

Figure 12. Proportional pressure control valve. (Reproduced by permission of Chapman and Hall.)

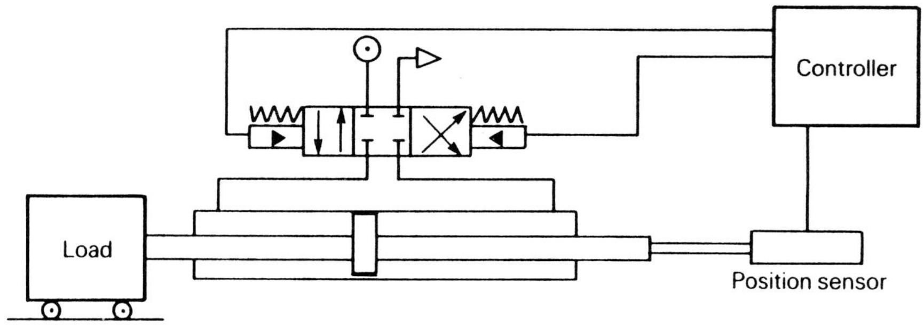

Figure 13. Position control of pneumatic cylinder. (Reproduced by permission of Chapman and Hall.)

selected, to be mechanically held. The addition of viscous damping using a self-contained hydraulic strut-type damper or vane damper may also be beneficial.

Systems such as that shown in figure 13 have been modelled using advanced digital techniques and implemented in hardware using a digital RIDE (Robust Inverse Dynamics Estimator) controller to control the valve [4]. This contains subroutines written in ESL (European Simulation Language) to model the valve orifices, treated as a number of discrete steps, the ram and load dynamics and takes into account Coulomb friction. Factors such as the saturation of the controller are also included.

An alternative to the spool-type proportional valve is to use pulse width modulation (PWM) of a poppet valve (figure 14). The valve is continually cycled open/shut, the amount of time for which it is open being determined by the mark/space ratio of the input signal. The average flow rate through the valve, at a constant pressure difference (never realized in practice!) is thus given by:

$$Q_{avg} = (T_1/T)Q_{max}.$$

At higher frequencies the valve can be held in a 'dither' mode at some intermediate position, depending on the inertia of the moving parts and the rating of the return spring. A limitation is the rise/fall time of the coil current and thermal effects will also need to be taken into consideration. The volume of air contained in the pipework and in the cylinder itself has an integrating effect as each slug of air is passed through the valve, the high

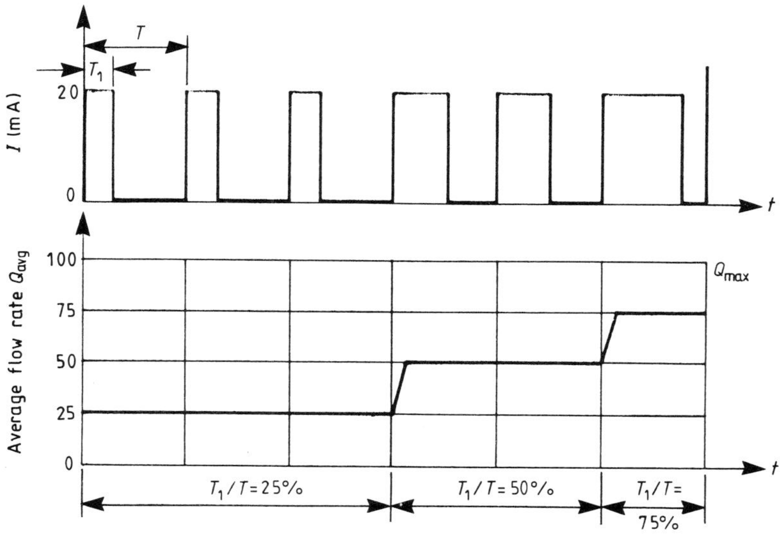

Figure 14. PWM of poppet valve. (Reproduced by permission of Chapman and Hall.)

compressibility and low density of air make PWM possible with pneumatics in a way which would be precluded in hydraulic systems because of the inevitable problems of surge and 'oil-hammer' effects.

Force-controlled pneumatic actuators Some of the most appealing potential applications of pneumatic actuators are in vehicles, particularly in suspensions. Here the requirements are of high volumetric flow rate and rapid response in applications which demand control of *force* rather than position. On commercial vehicles, air-operated brakes have been standard for many years but now require the addition of anti-lock braking system (ABS) features whereby the wheels are momentarily released if the wheel speed sensor detects a rate of angular retardation which indicates that the wheel is about to 'lock-up'. The valves used in such systems are usually two-stage in which a responsive small first stage is used to unbalance a much larger poppet or plate valve, thus permitting a large flow rate which can be maintained over a sufficient number of on/off cycles by means of an air reservoir mounted as close as possible to the actuator.

Other applications on vehicles include air-operated clutches, used in conjunction with pneumatically actuated gearshifts. Here, the requirement is for a very sensitive control of force, replicating the action of the driver's foot. Pressure control valves as in figure 12 have been used for this purpose.

The most demanding of vehicle applications is undoubtably suspensions, whether primary wheel-to-body or, in the case of heavy commercial vehicles, chassis-to-cab. The actuator in this case is a rolling diaphragm air bag constructed in reinforced rubber in such a way as to have high circumferential (hoop) strength but of itself negligible load-carrying capacity in the axial direction. When charged with air at the normal working pressure of 5 to 7 bar, it may be axially compressed with only a small change in cross-sectional area resulting, thus behaving similarly to a rigid-walled metal cylinder.

Here a number of control scenarios may be envisaged.

Quasistatic: vehicle ride height maintenance

This feature provides a constant ride height against changes in vehicle payload. Traditionally, a simple lever-action, mechanically operated valve has been used, but recent developments in, for example 'kneeling' buses for use by infirm or handicapped people, or flat-bed vans or trucks which need height adjustment to mate with supermarket loading bays, have called for the use of valves which can be remotely operated by the driver and these may be of the proportional type.

Intermediate response: vehicle roll control

Handling manoeuvres with commercial vehicles, particularly the reversal of steering angle conditions required in traversing a roundabout, sometimes result in alarming angles of roll. Air suspensions are worse in this respect than the steel multileaf springs traditionally used, which in practice are stiffer, have higher intrinsic damping, and which became stiffer as the interleaf friction builds up due to lack of lubrication and the ingress of dirt. The motivation for using air suspensions is that they provide a much more predictable and comfortable ride for passengers and load and also assist in reducing damage to the road itself.

Because of the present expense of providing an independent means of measuring lateral acceleration as the vehicle undertakes a manoeuvre, the normal practice is to derive lateral acceleration from the steering wheel angle and the road speed, possibly in conjunction with measurements of the relative displacement of the axle and body on opposite sides of the vehicle.

A circuit as in figure 15 may be employed using fast-acting proportional or poppet valves. Flow rates of up to 400 $1\ \mathrm{min}^{-1}$ at 5 bar pressure will be required for this service, with a system time constant of 200 to 100 ms. It is therefore evident that the valves which are adequate only for the quasistatic mode of control as described above will not meet this duty, but that the fast-acting poppet valves will normally be able to handle both regimes of control.

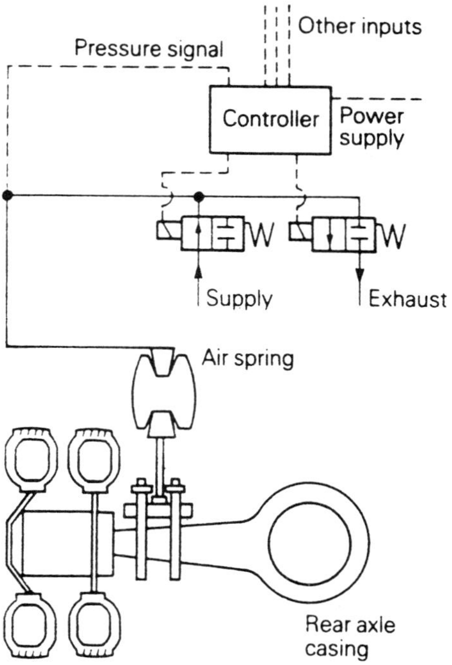

Figure 15. Air suspension with fast poppet valves. (Reproduced by permission of Chapman and Hall.)

Dynamic: control of vehicle ride quality
Many vehicles, usually passenger cars, now have driver-selected options for altering the characteristics of the primary suspension between 'city' and 'sports' or more descriptively, between soft and hard. This is usually accomplished by operating valves within the normal hydraulic shock absorbers in order to change the damping rate, since varying the rate of the traditional metal coil springs is impracticable. Air suspensions do however afford the opportunity of providing a form of active suspension control in that the effective spring rate may be varied by controlling the pressure in the air spring. If this can be done sufficiently rapidly, within the rise time of a single bump, then, provided that the suspension has sufficient physical working space, the suspension may be momentarily softened to allow a greater wheel movement and prevent the transmission of large forces (and hence uncomfortable accelerations) into the vehicle body. The drawback with all active suspensions systems is the additional power burden which this places on the vehicle systems and ultimately on the engine, which has adverse effects on the fuel consumption. A method which has been proposed for accomplishing this without the large power demands of the fully active system is the twin-chamber system (figure 16 [5]), which provides a means of altering the effective stiffness of the suspension in a dynamic fashion and is also regenerative in that the energy stored in the secondary (rigid) chamber is released into the air spring during the rebound. The vehicle air compressor therefore has to do no more than make up system leakage. This duty calls for a very large instantaneous flow rate and short response time, up to 1000 $\mathrm{l\,min^{-1}}$ and a response time of under 10 ms. This is outside the scope of commercially available valves, at least those which are sufficiently compact and inexpensive.

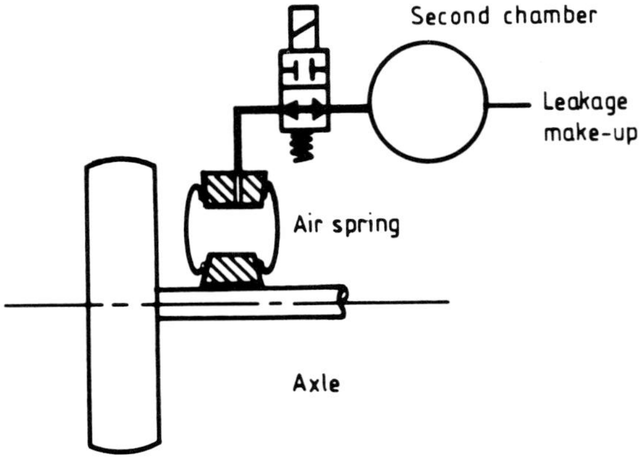

Figure 16. Twin-chamber air suspension.

The whole thrust of air valve development is towards achieving high flow, fast response and low cost. The limitations have been primarily in the first stage, due to the insufficiently rapid response of the solenoids commonly used. The use of rare-earth cores and pressure balance techniques has produced some notable improvements as in the valve of figure 17 which has the following operating parameters:

Rise time to open:	3 ms
Power input:	2.4 W
Flow rate at 8 bar:	300 $l\,m^{-1}$
Overall dimensions:	height 63 mm, plan 15 mm × 24.5 mm.

An even faster valve is the piezoelectric device (figure 18), which ingeniously employs an analogy of the bimetallic strip using bonded strips of piezoelectric material. This can produce a useful range of movement sufficient to operate a first-stage flapper orifice valve, with a power input of a few milliwatts.

It is the vehicle market, and the requirement for controllable suspensions in particular, which has also set in train a succession of developments in silicon-based accelerometers and position transducers reducing their cost by between one and two orders of magnitude from laboratory-standard instrumentation.

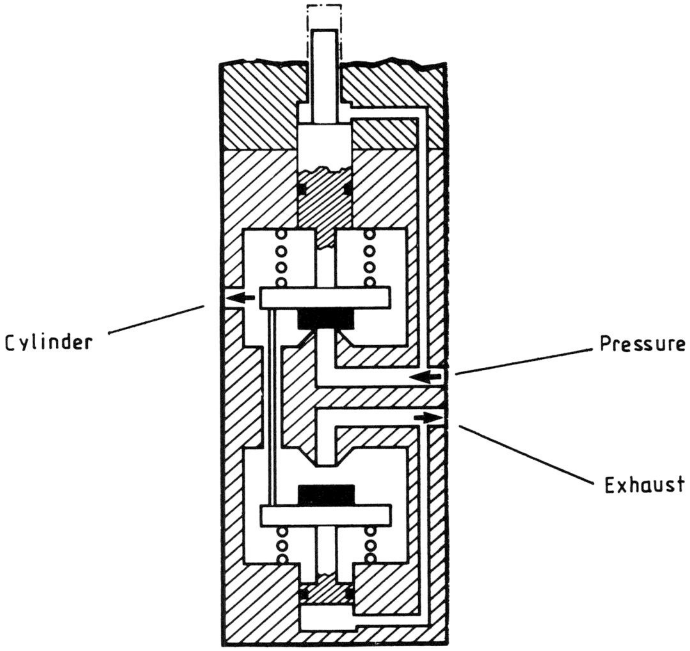

Figure 17. High-response double-poppet valve (Kuhnke).

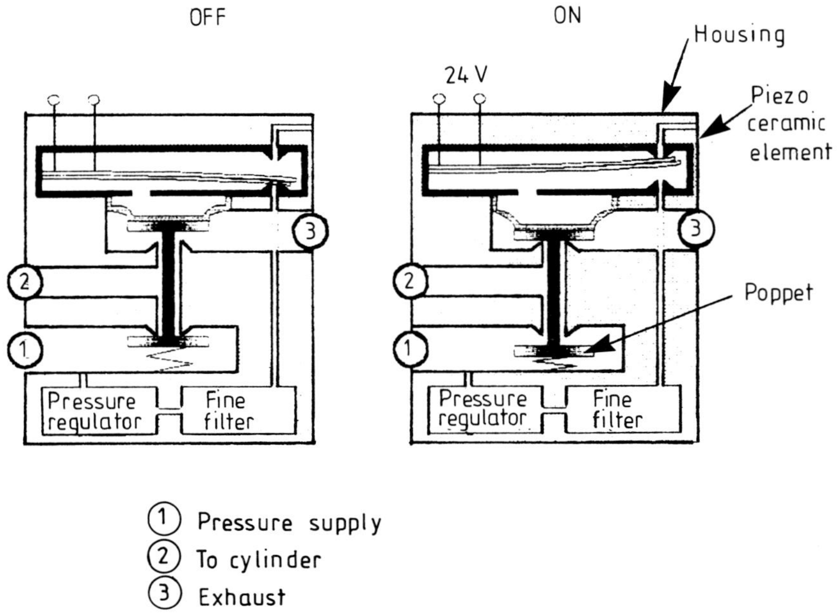

Figure 18. Valve with piezoelectric first stage (Hoerbigar).

2.4.5 Other Developments in Pneumatic Actuators

Systems which avoid the use of electrical signals in hazardous areas, or which are intrinsically proof against electromagnetic interference, continue to arouse interest. One such is the opto–pneumatic interface which allows direct connection of a fibre-optic cable to a pneumatic system [6]. Here, use is made of the minute quantities of thermal energy available in the light signal to create distortion in a diaphragm capsule and so create a back pressure across a control orifice which can then form the first stage of a fluidic amplifier.

2.5 HYDRAULIC ACTUATORS

2.5.1 Linear Hydraulic Actuators

These are typified by the familiar hydraulic cylinder, whose construction is shown in figure 19. Working pressures are between 100 and 300 bar (10–30 MPa) and may be higher in military equipment where space is at a premium. Hydraulic cylinders are almost always double-acting, but account must be taken of the fact that the forces available for a given pressure in the

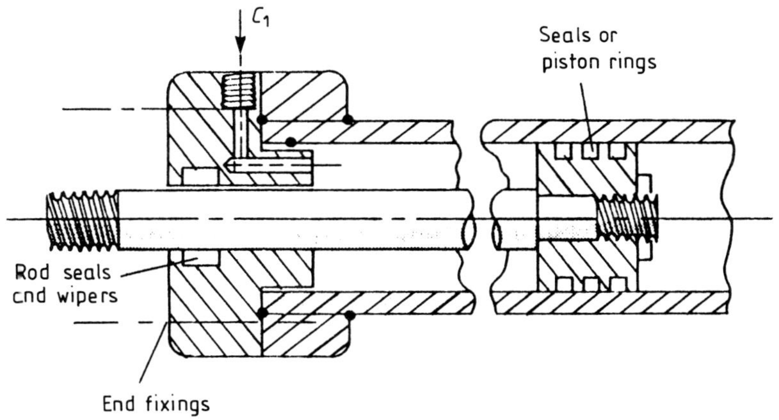

Figure 19. Construction of hydraulic cylinder. (Reproduced by permission of Chapman and Hall.)

two directions are very different. Because of the high forces available the rams have to be stocky to avoid Euler strut instability and lateral deflections which would damage the seals. Hence the annulus (rod end) area is about half the piston area and the 'retract' force is less than the extend force in proportion.

The motivation for using hydraulic cylinders is found in some combination of the following.

(1) Very high forces may be exerted in a compact format.

(2) Hold on load (hydraulic locking) can be obtained if the piston and valve seals are leaktight.

(3) With a suitable control system, very precise control of movement and position can be achieved.

(4) Overload protection is readily provided by means of a pressure relief valve.

It must be pointed out, however, that actuators based on electric motor/ballscrew systems are now competitive with hydraulic systems in several of these respects, apart from the last where it is hard to imagine a motor/ballscrew being back-driven by the load without serious damage resulting.

2.5.2 The Control of Linear Hydraulic Actuators

The majority of hydraulic applications are controlled by directional control valves which operate in a number of discrete positions giving full-flow/zero-flow/full-flow, otherwise known as 'bang-bang' mode. These are adequate for the driving of many industrial systems and may be manually operated, or if solenoid driven, may be readily interfaced to a programmable logic controller.

More demanding applications such as operating aircraft control surfaces or machine tool position control systems require the use of electrohydraulic proportional or servovalves. Both of these classes of valve employ a spool element as shown in figure 20. The basic flow equation for a spool element is given by

$$Q = 2\pi r C_d \left(\frac{P_s - P_1}{\rho} \right)^{-2}$$

where Q = volumetric flow rate ($m^3\ s^{-1}$), C_d = orifice discharge coefficient,

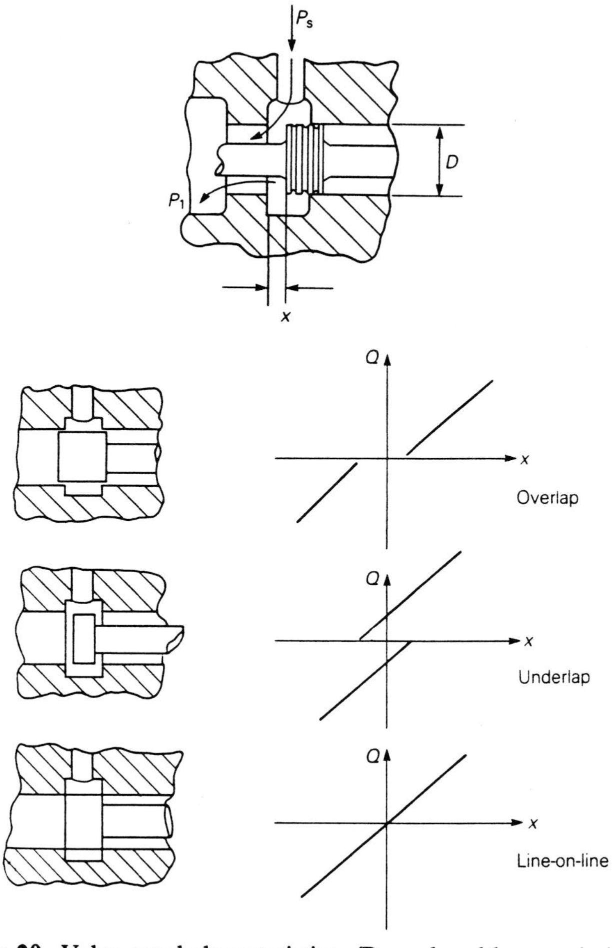

Figure 20. Valve spool characteristics. (Reproduced by permission of Chapman and Hall.)

r = spool radius (m), ρ = fluid density (kg m^{-3}) and P_s, P_1 = supply and cylinder pressure.

For a sharp-edged orifice, the flow rate is independent of viscosity and this is a fair assumption for the annular orifices found in spools. In a proportional directional control valve, the spool is configured to provide two control orifices as shown in figure 21. This illustrates a typical electro-hydraulic proportional valve with spring centring in which the offset, and therefore the flow rate at constant pressure, is proportional to the current in the solenoid which is energized. The principal difference between a proportional valve and a servovalve is that a servovalve usually incorporates a position transducer on the spool itself in order to control more effectively the position of the spool. A servovalve is normally zero-lapped, whereas a proportional valve is overlapped and thus has significant deadband.

Because servovalve spools have no elastomeric seals, they do not give a positive shut off, and unless cylinder lock valves are provided, the load may gradually slump when power is turned off. This has obvious safety implications.

A very important class of electrohydraulic servovalve is shown in figure 22 and is widely known as the Moog valve after its inventor who in 1947 created a high-performance valve design after examining the simple flapper orifice valves used to control the efflux vanes in the V2 rocket. The valve in its progressive refinements and through licensees has formed the basis for most aircraft flight controls since that time. In its commonest form, a torque motor and flapper orifice are used to create an unbalanced force on the valve spool. Equilibrium is restored in the controlled offset position of the spool when the torque, due to the bending of the 'feedback wire', is balanced by the torque exerted by the armature. The operating parameters of this type of

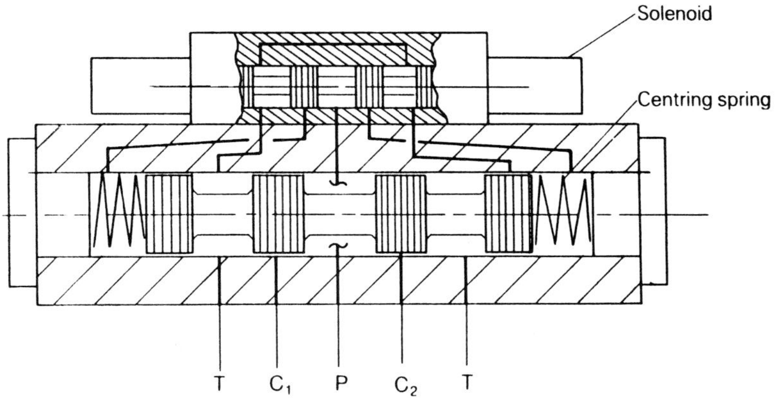

Figure 21. $\frac{4}{3}$ directional control valve. (Reproduced by permission of Chapman and Hall.)

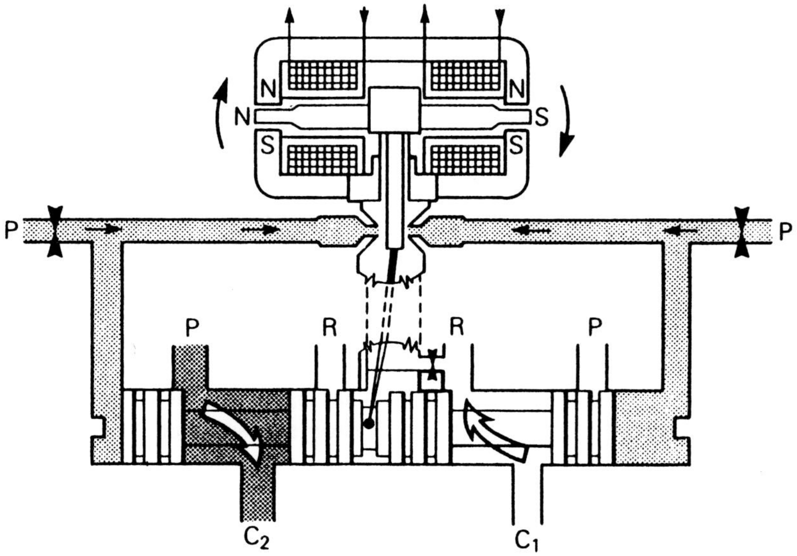

Figure 22. Electrohydraulic servo valve (Moog). (Reproduced by permission of Chapman and Hall.)

valve are somewhat higher than for the solenoid operated valve, and the cost is typically twice as great. It is claimed that this type of valve is also less prone to jamming, though this is controversial. Typical operating parameters are:

Input signal	0–40 mA DC at 12 or 24 V
Output flow	Up to 10 l s^{-1}
Working pressure	Up to 400 bar (40 MPa)
Linearity	Better than 1%
Frequency response	−3 dB roll-off not until 150 Hz to 500 Hz.

Moog-type valves are therefore used for actuation applications demanding high power and very good frequency response such as vehicle suspension test rigs, or indeed for the active control of vehicle hydraulic suspensions on the road. A history of the development of these valves is given in [7], and a fuller treatment of their operating characteristics with typical transfer functions in [8].

Hydraulic actuators are now readily available as packaged units, with a position transducer built into the ram and a servovalve mounted direct to the cylinder. In some recent units, the complete signal conditioning and power electronics package is mounted within the valve casing. Control functions such as proportional-integral-derivative (PID) algorithms and cycle repetitions can be made resident in hardware and in externally reprogrammable software within the valve housing itself.

With the responsiveness available from electrohydraulic servovalves, the control strategies which may be adopted are now very wide-ranging and many interesting approaches beyond the conventional PID algorithms have now been implemented [9].

2.5.3 Some Applications of Servo-controlled Linear Hydraulic Actuators

A number of industrial robots have employed hydraulic actuators, but these have now been displaced by high-torque electrical drives. However, in some larger 'robot-like' machines such as steel billet manipulators and construction equipment, servo-controlled hydraulic actuators are making an important contribution. An example is shown in figure 23 of a $\frac{1}{5}$ scale robotic back-hoe whose control system is now implemented on a full-size machine. The objective is to provide driver aids, such as the ability to scrape the bottom of a trench at constant depth or to a controlled fall with the manipulation of one lever only. Eventually the aim is to progress to a fully autonomous machine which could work, for example, in a contaminated environment without a human operator being present. The back-hoe has a hydraulically driven slew function and three software coordinated rams

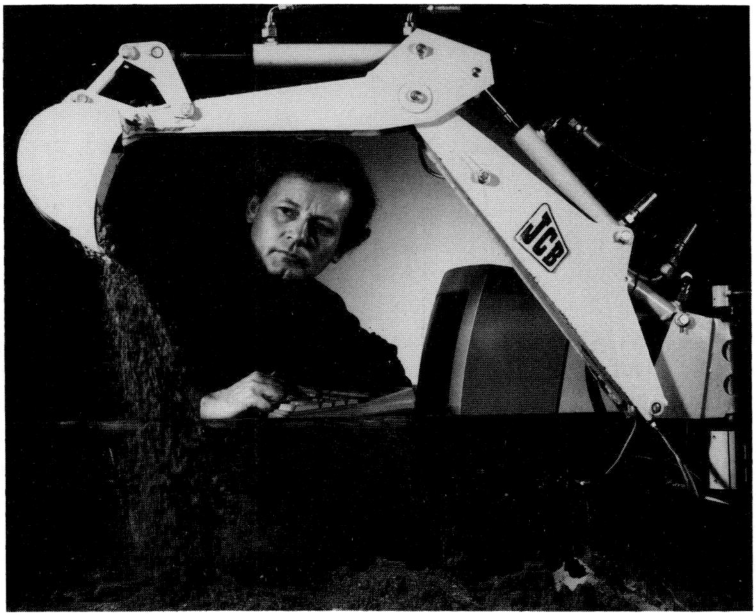

Figure 23. Working $\frac{1}{5}$ scale model of back hoe.

controlling the arm and bucket. Each ram is controlled by a solenoid-driven servovalve, the dynamic response of this type being quite adequate for the purpose.

A more demanding application dynamically is the active suspension in which acceleration and position feedback is used to control a hydraulic ram at each wheel. In the practical systems now being developed, the duty of a controlled hydraulic strut is simplified by employing it in parallel with a conventional passive metal spring which supports the static weight of the vehicle. High-response flapper-controlled servovalves are currently employed, but once again the penetration of such beneficial developments into the large-volume automotive market will in due course result in the production of application-specific valves at much reduced cost.

Diesel engine fuel pumps and injectors, although not hydraulic actuators as such, do have some features in common. The injection process and its timing involves the transmission of large forces which have made it difficult to bring in a 'mechatronic' approach. However, the benefits to be conferred by on-line injection timing and calibration are so great that it would not be surprising to see radically new concepts perhaps involving piezoelectric devices being brought into this application within the next few years.

2.5.4 Rotational Hydraulic Actuators

Hydraulic motors form the greater part of this class of machine, although there are limited angle actuators of the vane type resembling the pneumatic actuators of figure 5. The variety of hydraulic motors available is very wide indeed, but all are reversible in the sense that they can function equally well as a pump or as a motor. All such machines are hydrostatic in operation and have a specific displacement which is normally expressed in cm^3 per revolution, or in similar volume/angular terms. However, this is an *ideal* or geometric displacement which, in the case of a piston pump, is a function of such parameters as the number of cylinders, their bore and stroke.

All hydraulic motors are subject to *internal* leakage due to the passage of high-pressure hydraulic oil through the working clearances between the pistons (if such is the type) and the bores as no seals are used. This leakage is put to natural good effect in lubricating the mechanical parts within the machine casing, but requires a third pipe (in addition to the high-pressure feed and return) to take it away to tank. The principal effect of internal leakage is to make the actual displacement less than the ideal. Since the output speed is proportional to the volumetric flow rate, it follows that the output speed is less than would be expected from the geometric displacement. This is expressed as a hydraulic efficiency:

$$\mu_H = N_{actual}/N_{ideal}.$$

Axial piston machines are most efficient at about 96%, but simple gear motors may have efficiencies as low as 70%. As a general rule, efficiency tends to be lower with reduced overall equipment size, lower speeds and higher pressures. In manufacturers' literature, efficiency is shown on performance charts as in figure 24.

Hydraulic motors are divided into two broad categories: low speed, as exemplified by the radial piston machine in figure 25, and high speed, as in the axial piston machine (figure 1).

The radial piston machine finds much use in the process industry for the drive of chemical mixers, rubber mills and cement kilns. It is also found in direct wheel drives in agricultural and forestry vehicles and in the winch drum drives of cranes.

A particularly advantageous feature of low-speed high-torque hydraulic motors over geared electric motors is that they have a much lower kinetic energy of rotation. For example, consider a process machine running at 50 rev/min directly connected to a hydraulic motor. An electrical drive would employ a 30:1 reduction gearbox and a 1450 rev/min electrical motor. In a 'jamming' situation, the gearboxes in electrical drives have been known to disintegrate, due to the residual kinetic energy of the motor after tripping, having been in excess of the strain energy which the downstream parts of the mechanical system could absorb without breakage. As is well known, mechanical torque limiters are prone to interference and tend not to work when they are needed! A further advantage of this type of machine is that they are almost silent in operation.

The common high-speed axial piston motor/pump runs at electric motor speeds, generally in the range 1000 to 4000 rev/min. They are more

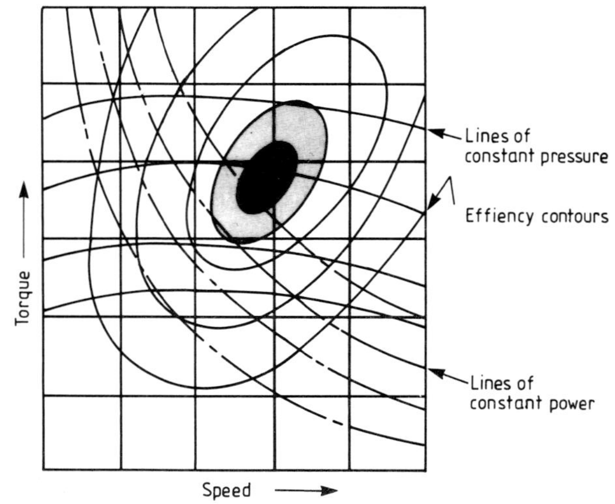

Figure 24. Form of performance charts for hydraulic machines.

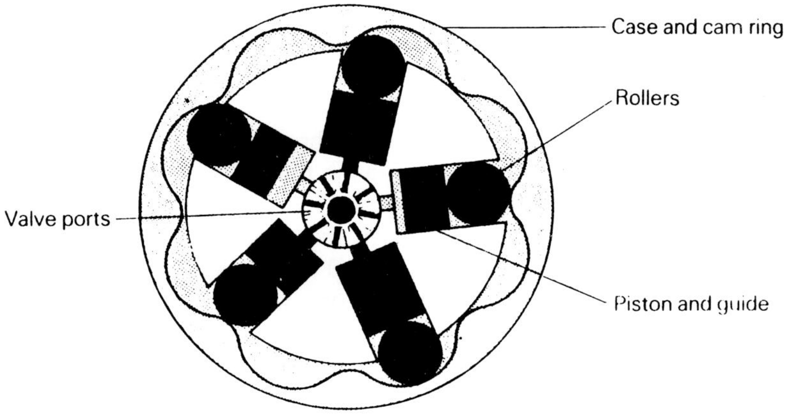

Figure 25. Radial piston hydraulic motor. (Reproduced by permission of Chapman and Hall.)

commonly used as pumps to supply a hydraulic circuit, in which case the *variable displacement* facility provided by the variable angle swashplate is particularly useful. Referring to figure 26, the torque and volumetric flow rate for an axial piston machine are given by:

$$T = CpaR \tan \theta$$

where T = torque (N m), $C = 2.15$ approx, p = system pressure (N m^{-2}), a = area of one piston (m^2), R = pitch radius of cylinder barrel (m), θ = swashplate angle (radians),

$$Q = fNaR \tan \theta$$

where Q = volumetric flow (m^3 s^{-1}), f = rotational frequency rad (s^{-1}), N = number of cylinders,

$$D = NaR \tan \theta$$

where D = specific displacement (m^3 rad^{-1}).

The torque in such a machine is not constant, but undergoes a small fluctuation as each piston crosses the boundary between the suction and discharge side of the valve plate. For example, in a machine with seven pistons, the factor C fluctuates between 2.11 and 2.18 seven times per revolution. The valving action which involves some small compression of the high bulk modulus fluid is also the principal source of noise in hydraulic systems, and measures to alleviate this at source usually pay a small price in reduced efficiency.

A typical implementation of a rotational hydraulic actuator (or drive) is shown in figure 27. This is a hydraulic transmission with speed control which could be employed, for example, to drive a process mixer. In this

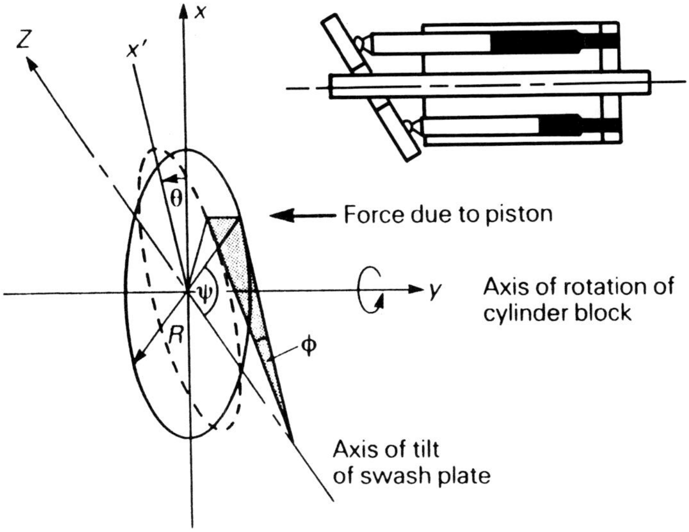

Figure 26. Geometry of axial piston machine.

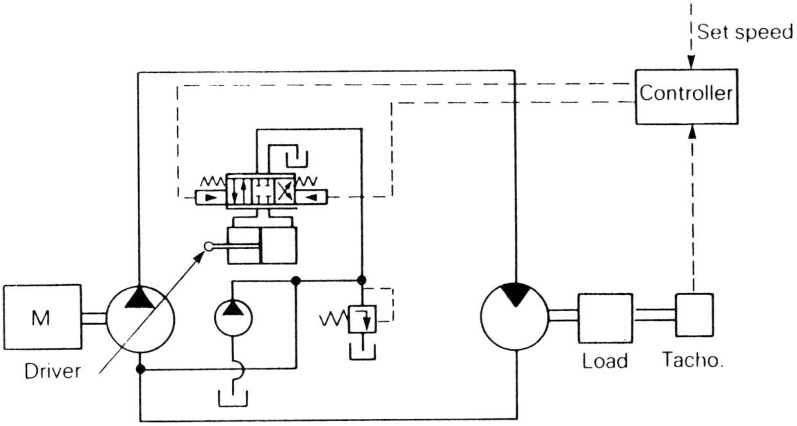

Figure 27. Variable pump in speed control. (Reproduced by permission of Chapman and Hall.)

case, the pump would be an axial piston machine and the motor a low-speed radial piston type. The specific displacement of the motor would be much greater than the nominal maximum of the pump and thus the speed and torque ratios would be given by:

$$N_2/N_1 = D_1/D_2 \qquad \text{and} \qquad T_2/T_1 = D_2/D_1.$$

These are the ideal ratios, assuming that there is no internal leakage, but in practice the output torque is reduced by about 2% to 4% due to mechanical losses, and the speed by a larger proportion due to internal leakage. The actual output ratio would therefore be:

$$N_2/N_1 = (\mu_{H_M}/\mu_{H_P})(D_1/D_2).$$

A hydraulic transmission therefore has a torque ratio analogous to a mechanical gearbox, but is not in itself a positive drive. Synchronization between input and output cannot be maintained without monitoring the shaft position. If both pump and motor have a variable displacement capability, then a wide range of output characteristics can be achieved.

The load characteristics must be specified in terms of load/speed before the drive can be selected and for rotational drives these can be divided into three types:

(1) constant torque—exemplified by winching and craneage operations;

(2) constant power—accelerating heavy rotors (e.g. centrifuges) and vehicles;

(3) drives in which: $T = CN^2$—typical of hydrodynamic machines, pumps, fans and mixers.

In practice, systems may not be represented by these idealized cases and external disturbances and non-linear effects will need to be taken into account.

Many industrial hydraulic rotational drives are of the type shown in figure 27, but various, mostly small-scale, grouped drive installations exist in which a central pumping unit serves a number of hydraulic motors. The common supply main is held at constant pressure using a pressure compensated pump in which the volumetric flow output is modulated in accordance with the system's total demand. The simple hydromechanical swashplate controllers are reliable in operation and can vary the output from zero to maximum flow in about 20 ms. The implementation of a parallel connected hydraulic ring circuit is shown in figure 28. Speed control of the individual

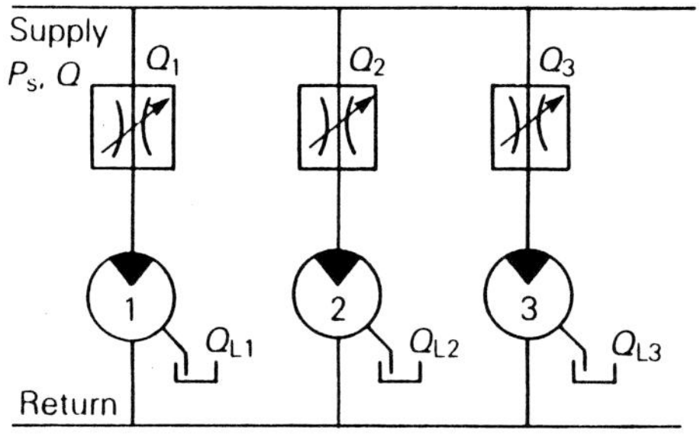

Figure 28. Hydraulic motors on common main. (Reproduced by permission of Chapman and Hall.)

hydraulic motors is achieved by means of a pressure-compensated flow control valve which regulates the flow rate to the motor. The valve compensates for pressure variations which occur at the inlet to the hydraulic motor due to changes in the load. Such valves are very *effective* in operation but are not *efficient* in that they introduce a pressure drop which may be very large. Under adverse operating conditions, 50% or more of the input power to the system may be rejected as heat to the working fluid.

Various means have been proposed to ameliorate this problem including self-tuning pressure control. This involves monitoring the conditions at each driven machine using torque and speed-sensing, and modulating the system pressure to just match the pressure requirement of the most heavily loaded unit. Investigations have been made [10] of the use of variable displacement hydraulic motors in systems, such as that of figure 29 where the torque output, at constant pressure, of the hydraulic motor can be matched to the load over a wide speed range. At the present time, only high-speed machines offer the facility of fully proportional variable displacement, but Chapple [11] gives details of a high-torque motor which demonstrated this capability. To the author's knowledge, this promising line of hydraulic system control awaits its first industrial application.

Further developments in hydraulic pump/motors [12] have demonstrated the use of magnetically latched poppet valves replacing the traditional ported valve plate on axial piston machines. The valves may be individually timed to provide a variable displacement facility with very high rates of response, and by eliminating the need for a variable swash angle the design has provided a mechatronic solution in simplifying the mechanism.

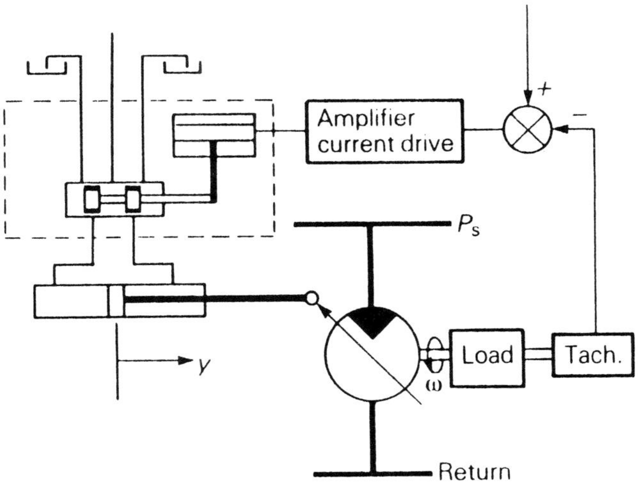

Figure 29. Variable displacement motor with speed control on common main. (Reproduced by permission of Chapman and Hall.)

ACKNOWLEDGMENT

The kind permission of Chapman and Hall is acknowledged for the use of figures 1, 3, 4, 5, 6, 7, 12, 13, 15, 19, 20, 21, 25, 27, 28 and 29 from the book *Mechatronics—Electronics in Products and Processes* by D A Bradley *et al* (1991).

REFERENCES

[1] Douglas J F, Gasiorek J M and Swaffield J A 1985 *Fluid Mechanics* (Boston, MA: Pitman) pp 613–629
[2] Bracewell R H, Chaplin R V and Bradley D A 1993 Schemebuilder and layout—computer-based tools to aid in the design of mechatronic systems *I. Mech. E Conf. on Mechatronics—The Integration of Engineering Design (Dundee, 1992)* pp 1–7
[3] Esposito A 1994 *Fluid Power with Applications* (Englewood Cliffs, NJ: Prentice Hall) 3rd edn pp 435–452
[4] Bradshaw A and Counsell J M 1993 Advanced digital controller for servo-pneumatic actuated machines *I. Mech. E Seminar on Machine Actuators (London, 1993)* pp 5–9
[5] Dawson D 1988 Adaptive suspension for a medium weight truck *Proc. 18th ISATA (Florence, 1988)* pp 382–387
[6] Landry M J, Rupert J W and Mittas A 1992 Power-by-light systems and their components—an evaluation *Appl. Optics* **30** 1052–1061
[7] Maskrey R H and Thayer W J 1978 A brief history of electrohydraulic servo-mechanisms (*Moog Inc. Controls Div. ASME Dynamic Systems Measurement and Control*)
[8] Thayer W J Transfer functions for Moog servovalves *Technical Bulletin* **103** (Moog Inc. Controls Div.)
[9] Bradshaw A and Counsell J M 1992 Design of autopilots for high performance missiles *Proc. I. Mech. E. J. Systems and Control* **206** pp 75–84
[10] Lange J 1987 Control of variable displacement hydraulic motors supplied from constant pressure ring mains *MPhil Thesis* Lancaster University
[11] Chapple P J 1982 *I. Mech. E* Bramah Medal Presentation
[12] Rampen W H S and Salter S H 1990 The digital displacement hydraulic piston pump *Proc. 9th Int. Symp. Fluid Power BHR Group (Cambridge, 1990)* pp 33–36

3

Electro-rheological Actuators

Roger Stanway and John L Sproston

When compared with actuators powered by electrical or pneumatic energy sources, hydraulic actuators can offer significant advantages in terms of power-to-weight and force-to-mass ratios.

Unfortunately the servovalves which control the pressure/flow characteristics in many hydraulic systems are expensive to produce and contribute to a relatively high overall cost. If the cost of devices for controlling fluid flow could be reduced whilst maintaining the performance of a hydraulic system then this type of actuator would find many more applications in low-cost automation schemes.

In this chapter, the authors describe one technique for reducing the cost of hydraulically powered actuators using so-called 'electro-rheological' (ER) fluids. The resistance to flow of ER fluids can be controlled through an applied electric field and this leads to the concept of a flow control valve with no moving parts. After describing the basic properties of ER fluids, the principles of operation, construction and performance of ER valves are discussed. The manner in which ER valves are employed in hydraulic circuits for force actuation is then described, and the performance of typical ER actuators is summarized. To place the ER actuator in perspective, some of the factors which have prevented more widespread exploitation are explained and finally some current developments, aimed at overcoming these difficulties, are outlined.

3.1 INTRODUCTION

Of the three main types of supporting energy (electrical, pneumatic, hydraulic) which are candidates to provide power for modern actuators, hydraulic energy is generally favoured where power-to-weight ratio is of prime importance. Compared with electrical generators and motors, Burrows (1972) suggests that equivalent hydraulic pumps and motors offer a power-to-weight advantage of around ten-to-one. In providing rectilinear motion, the force-to-mass ratio available from hydraulic units is liable to be one hundred times better than an electrically driven system. Moreover, when connected to a pressurized source, a hydraulic actuator possesses an inherent output stiffness which is not present in an electrical system without the use of feedback. Because of these advantages, hydraulically powered actuators have found a wide range of applications in such diverse fields as aerospace vehicles, earth-moving equipment, and robotics for manufacturing processes.

An actuation system which uses hydraulic energy generally contains three main components—the pump, the valve and the actuator (or motor) itself. The valve, which modulates the flow of fluid from the pump to the actuator, is a complex piece of equipment which accounts for a large portion of the overall system cost. If the complexity and cost of the valve could be reduced, then hydraulic actuators would find an even wider range of applications, especially in low-cost automation schemes.

A closer look at the conventional hydraulic valve reveals that it consists of three main subsystems—a torque motor, an hydraulic pre-amplifier and a control spool (Morse 1963). Together, these provide the necessary amplification but also contribute to the relatively high cost of hydraulic valves as each subsystem contains components which must be manufactured to very fine tolerances. To obtain a significant reduction in cost requires a drastic reduction in the total number of components, preferably combined with a relaxation of tolerances.

In this chapter, the authors describe one promising strategy for reducing the cost of hydraulically powered actuators. This strategy involves the replacement of conventional hydraulic fluid by so-called ‘electro-rheological’ (ER) fluid. Such special fluids consist of a suspension of semiconducting particles in a dielectric carrier liquid. Upon the application of an electric field of sufficient intensity a significant (and reversible) change in resistance to flow occurs in the fluid. Using this principle, the flow of an ER fluid can be modulated through the applied electric field and thus the flow from a pump to an actuator can be controlled by what is effectively an ER valve.

Unlike its conventional counterpart, an ER valve has no moving parts and the tolerances are no longer as critical. There is little doubt that the manufacturing cost of an ER valve is significantly less than that of a hydraulic device. However the question arises: can an ER-based actuation system match the performance of a conventional hydraulic system? With emphasis on rectilinear actuators, this chapter describes the design and construction of ER valves and shows how these are used to construct actuation systems. The performance of typical systems is discussed together with factors which have prevented wider commercial exploitation. Future prospects for ER actuators are also discussed.

3.2 FLOW CONTROL USING ER FLUIDS

3.2.1 What is an ER Fluid?

An ER fluid consists of a dielectric carrier fluid into which are dispersed semiconducting particles. The sudden application of an electric field of sufficient intensity causes polarization of the particles which consequently arrange themselves into chains, thus forming mechanical bridges across the electrodes. Particle dimensions range (typically) from 0.04 μm to 50 μm with electrode gaps of between 0.25 mm and 2.0 mm. The formation of these chains, and their subsequent breakdown when the electric field is removed, can readily be observed under the microscope.

In macroscopic terms, the ER fluid undergoes a transition with increasing field strength. At low field strengths (below 1 kV mm^{-1}) the ER fluid is essentially in a liquid state, similar to that of the carrier fluid. At field strengths above 2 kV mm^{-1} the ER fluid appears to behave more like a solid as a significant number of particle chains will have formed. Simple experiments demonstrate that significant changes occur in the resistance to flow of the ER fluid as the field strength is increased over the range from zero to 4 kV mm^{-1}, beyond which point electrical breakdown is liable to

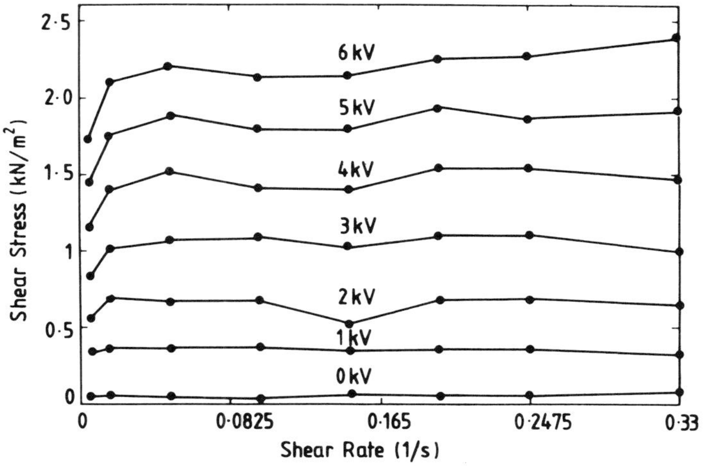

Figure 1. Shear stress versus shear rate characteristics for typical ER fluid in shear mode.

occur. Also the mechanical properties of the ER fluid in shear, tension and compression are subject to dramatic variations with applied field. It is often stated in the literature that ER fluids behave like a Bingham plastic in which the yield strength is a function of the applied field. The extent to which the mechanical stress/shear rate relationship can be varied is shown in figure 1, which refers to a typical ER fluid placed in shear (Steven 1985).

As one might expect, the greater the proportion of solid material in the composition of an ER fluid, the greater the values of yield strength which can be obtained for a given electric field. However, there is a price to be paid for improved mechanical performance in the form of increased electrical power consumption. Current density can be shown to increase dramatically with increasing solid content (Duclos, Acker and Carlson 1988). The key mechanical and electrical properties of ER fluids are summarized in table 1. The main chemical/physical properties are listed in table 2.

Table 1. Mechanical and electrical properties of ER fluids.

Mechanical properties	Yield stress	Yield stress = aE^b where exponent has been determined to lie in the range from 2.0 to 2.4. Increases with volume fraction and is temperature-dependent (Jordan and Shaw 1989)
	Steady shear viscosity	Apparent viscosity $\propto E^2$ and increases with volume fraction. Zero-field viscosity can be significant in practical devices, see Stevens *et al* (1988)
	Dynamic modulus	Viscoelastic property i.e. ER fluids produce in-phase (storage) and out-of-phase (loss) components in response to sinusoidal mechanical stimulus
Electrical properties	Conductivity	Non-ohmic and increases dramatically upon the addition of water. Temperature-dependence can lead to thermal runaway (Stangroom 1983)
	Field strength	ER fluids typically can sustain a maximum electric field of 4 kVmm^{-1} before breakdown occurs. The field strength may be steady, pulsed or alternating
	Dynamic permittivity	Frequency-dependence of dynamic permittivity suggests that ER mechanism may be based upon time-dependence of particle polarization

Table 2. Chemical/physical properties of ER fluids.

Chemical/physical properties		Essential features
Particles	Shape	Basically spheroidal but various shapes have been tried. Little published information (Jordan and Shaw 1989)
	Size	Generally ranges from 0.04 to 50 μm. Smaller particles tend to suffer from influence of Brownian motion, larger particles are slow to respond to electric field
	Material	Commonly used materials are silica, clay minerals, metal oxides and starch: all water activated. For description of anhydrous particles, see Block and Kelly (1987)
Dispersant	Desirable properties	Low conductivity, low viscosity, high boiling point, low freezing point, high electric strength. For general discussion of properties, see Stangroom (1983)
	Material	Insulating oils, for example, silicon oil. For comprehensive list see Block and Kelly (1988)
Suspension	Volume fraction	Typically ranges from 0.001 to 0.4. Storage component of dynamic modulus increases with volume fraction
	Water	Evidence to support the view that water strengthens particle interaction and may serve as vehicle in ionic polarization mechanism
	Surfactants	To improve stability of suspension

3.2.2 The ER Valve: Principles of Operation

The results shown in figure 1 were obtained by using an ER fluid in what is normally termed the *shear mode* of operation. This term implies that the fluid is sheared by the relative motion of a pair of electrodes whose surfaces remain parallel to each other. The shear mode is employed (typically) in clutches for torque transmission (Stevens, Sproston and Stanway 1988) and vibration damping devices (Stanway, Sproston and Stevens 1987).

An alternative mode of operation, and the one which is exploited in ER valves, is the *flow mode*. In the flow mode, ER fluid is pumped between a pair of electrodes and by controlling the applied electric field the pressure drop/fluid flow characteristics can be varied by a significant amount. In a conventional hydraulic valve, the pressure drop/fluid flow characteristics are varied by providing an electric current to the coil of a torque motor which, through a hydraulic amplifier (typically a nozzle flapper valve), drives the control spool. The displacement of the control spool governs the pressure/flow characteristics of the valve.

To explain the alternative concept of flow control in an ER valve, it is helpful to begin by considering a conventional Newtonian fluid, flowing between a pair of identical, rectangular plates. If the length of each plate is L, the width is W and the gap between the plates is H, then for a flow rate of Q the pressure drop, P_O, is obtained from classical fluid mechanics (for example, see Massey (1983)):

$$P_o = 12\mu \frac{QL}{H^3 W} \tag{1}$$

where μ is the *Newtonian* viscosity of the fluid.

If an ER fluid is pumped between the same pair of plates, then it is reasonable to suppose that in the absence of an electric field the carrier liquid will behave like a Newtonian fluid and give rise to the pressure drop predicted by equation (1). However, it has already been established that an ER fluid is capable of developing a yield stress when an electric field is applied. Thus the ER effect itself will produce a further pressure drop across the length of the two plates. If the relevant *yield* stress is denoted by τ_y, then this additional pressure drop, P_{ER}, is given by

$$P_{ER} = 2\tau_y \frac{L}{H}. \tag{2}$$

The relationship between the pressure drops given by equations (1) and (2), and the extent to which control over pressure/flow characteristics is available using ER fluids, is shown in figure 2. Referring to figure 2, the total pressure drop, P_{TOT}, is made up of two components. The first of these components, P_O, results from the viscous flow of the carrier fluid while the second component, P_{ER}, comes from the ER effect and is thus controllable through the applied electric field.

Unfortunately, while the relationships between pressure drop and flow shown in figure 2 are useful for visualizing flow control in ER devices, the use of equation (2) to model the ER component of pressure drop does not accurately reflect the actual physical situation. The term τ_y in equation (2) fails to account for the accepted fall in shear stress with increasing shear rate. Brooks (1989) suggested that the use of equation (2) can lead to errors of around 20% in estimates of the total pressure drop. To obtain a better model, it is now generally accepted that a more reliable approach involves treating the ER fluid as a 'plug' of unsheared fluid flowing in the valve.

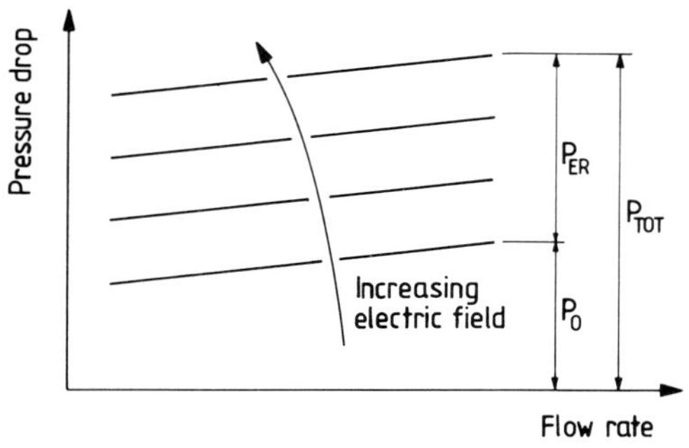

Figure 2. Relationship between pressure drop and flow rate for ER fluid in flow mode.

The plug flow assumption results in the so-called 'Bingham plastic' model (Phillips 1969) of an ER fluid which can be explained with reference to figure 3. Figure 3(*a*) shows a typical profile for one-dimensional fluid flow near to a fixed wall. The fluid velocity, u, is given by $u = \mathrm{d}x/\mathrm{d}t$ and thus, in the coordinate system of figure 3(*a*), the shear rate is given by $\mathrm{d}u/\mathrm{d}z = \mathrm{d}^2x/\mathrm{d}z\mathrm{d}t$. The total stress, τ_{TOT}, is then given by

$$\tau_{\mathrm{TOT}} = \tau_{\mathrm{y}} + \eta \frac{\mathrm{d}u}{\mathrm{d}z} \tag{3}$$

where η is defined as the *plastic* viscosity of the fluid flow which is assumed to be independent of electric field strength. The respective shear stress/shear rate characteristics for a Bingham plastic and for Newtonian fluid are shown in figure 3(*b*). The mechanism through which the Bingham plastic model of an ER fluid is applied to flow control between flat plates is considered in detail by Phillips (1969).

3.2.3 Construction of an ER Valve

In practice, rather than construct ER valves using flat electrodes, it is usually more convenient to build a valve as a series of coaxial cylinders. A typical arrangement is shown in figure 4 which shows a prototype configuration involving six coaxial cylinders. Alternate cylinders are connected to high tension and earth, thus establishing an electric field in the annulus between each adjacent pair of cylinders. The strength of the electric field governs the total yield stress of the fluid between adjacent cylinders and consequently the pressure drop across the device and flow of fluid through the annuli.

A design of valve suitable for commercial applications is illustrated by Brooks (1989). This device consists essentially of three components: an inner electrode assembly, an outer container which forms the earthed electrode and an end cap. Brooks noted that the design is elegant, simple,

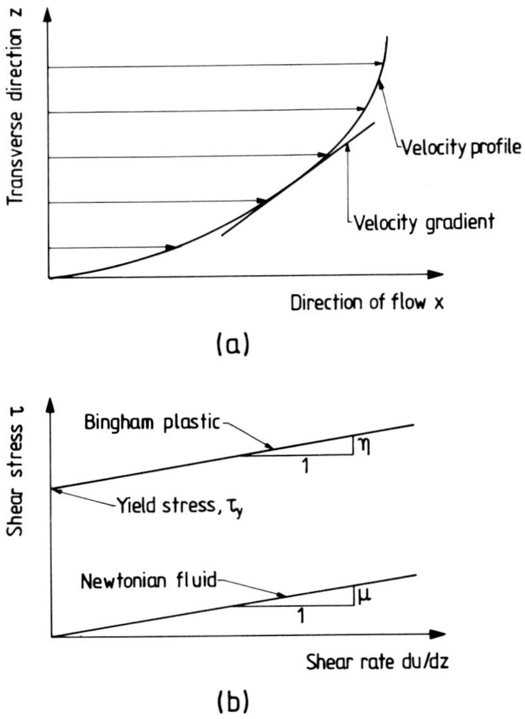

Figure 3. (*a*) Velocity profile and gradient for ER fluid flow close to a wall. (*b*) Shear stress versus shear rate characteristics for Bingham plastic and Newtonian fluid.

reliable and could be cleaned and changed with particular ease. It was also claimed that pressure drops of up to 60 bar have been produced by such valves of length no greater than 10 cm.

3.2.4 Performance of ER Valves

A typical arrangement for testing an ER valve consists simply of a hydraulic circuit in which fluid is pumped through an ER valve into a reservoir before being returned to the pump. Measurements of pressure drop and fluid flow are taken at different values of applied electric field.

For a model valve such as that shown in figure 4, the maximum pressure drop might be around 1 bar at a flow rate of 1.5 litres/minute with P_0 accounting for approximately 20% of the total pressure drop. Commercial valves are, however capable of much greater performance. To cite typical examples, Brooks (1982) quotes a maximum drop of around 30 bar with a P_0 of approximately 5 bar at a flow rate of 10 litres/minute with an electric

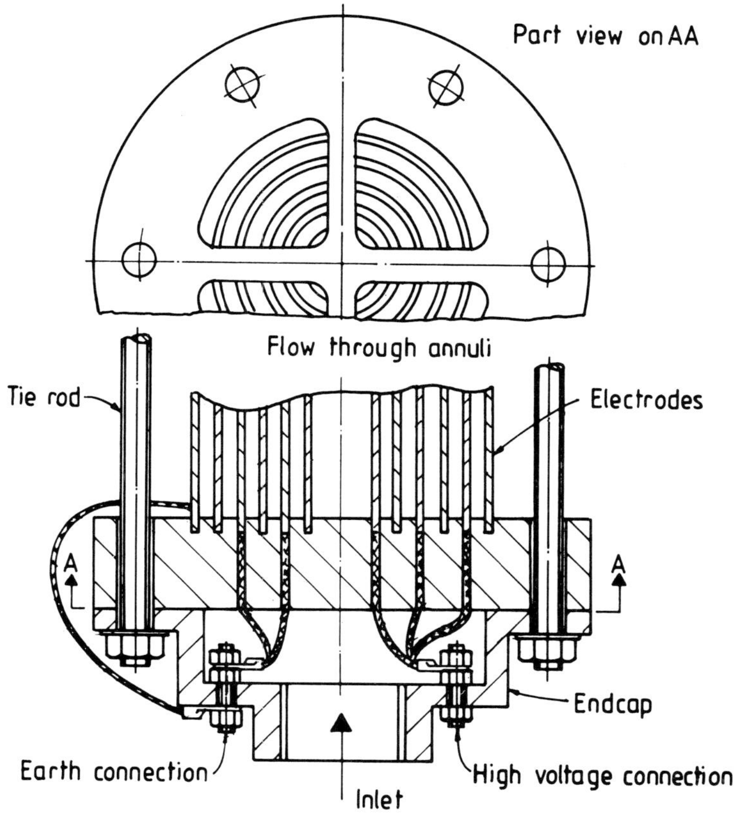

Figure 4. Model ER valve showing arrangement of coaxial cylinders.

field of 3 kV mm^{-1}. A later paper by Bullough (1988) shows a pressure/flow characteristic for a 'typical' ER valve where pressure drops of up to 70 bar are achieved at flow rates of no more than 2 litres/minute and again with an electric field strength of around 3 kV mm^{-1}.

From the figures quoted above, it is readily apparent that significant force levels are capable of being generated. In the section which follows, we consider ways in which the ER valve can be used as an integral part of an actuator system.

3.3 THE ER ACTUATOR SYSTEM

3.3.1 Hydraulic Circuits

A valve-controlled, hydraulic actuator normally consists of a pump, a number of flow-control valves and the motor or actuator itself. When operating under closed-loop control such an arrangement is usually called a hydraulic servomechanism (Morse 1963).

The hydraulic valves are often connected as a Wheatstone bridge which provides a balanced arrangement, providing efficient operation without the need for mechanical biasing of the actuator piston. In a similar way, four ER valves are usually arranged to form a balanced circuit, as shown in figure 5. The top and bottom junctions of the bridge are connected, respectively, to the fluid supply and exhaust. The two remaining junctions are connected to the actuator itself—effectively a piston translating inside a closed cylinder. By increasing the flow resistance on one side of the ram and reducing it on the other, the fluid flows into one side of the piston and produces a force which is a function of the pressure drop and piston area. The available speed of the actuator is a function of the flow rate into the chamber.

In the construction of hydraulic actuators, it is generally considered good practice to minimize the enclosed volume from the control spool to the actuator piston. This is often achieved by designing the manifolds of the valve directly onto the actuator body. The design and construction of ER actuators can also be approached in this way and a system in which the valves and actuators are integrated is described by Brooks (1982).

3.3.2 Performance of Typical ER Actuators

Probably the first report of the experimental performance of an ER actuator was that by Eige and Fraser (1961) who carried out investigations sponsored

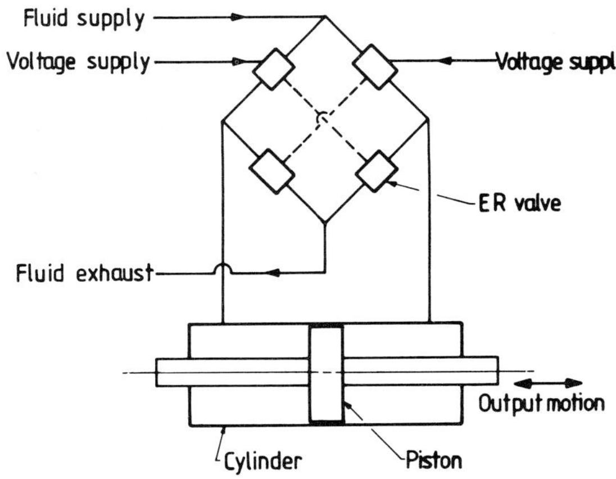

Figure 5. Hydraulic circuit for ER actuator showing Wheatstone bridge configuration.

by the United States Air Force. The apparatus described by these authors produced peak accelerations of a 1 kg mass of up to 30 m s^{-2} up to a frequency of approximately 200 Hz. However the tests were not considered a success as the specification required that this level of performance should be delivered up to a frequency of 2 kHz. Also, problems were encountered with the fluid itself which was thought to be a suspension of silica gel in a hydrocarbon or polysiloxane liquid.

A novel arrangement involving two pumps, each supplying an ER valve placed on either side of the actuator piston was described by Phillips (1969). In this arrangement the valves were excited electrically in such a way as to increase flow resistance on one side of the piston and reduce it on the other. Consequently the piston was displaced towards the valve with the lower pressure. Following a comprehensive computer simulation study, Phillips claimed that such a device was, in principle, capable of accelerating masses of 'several hundred pounds' (weight) at frequencies of up to 3.5 kHz. Unfortunately, no experimental results were presented and so it is not possible to test these claims.

Developments in ER fluid and device technology made possible significant improvements over the next decade and Brooks (1982) described the design and construction of a 'high-frequency, high-thrust' actuator. This unit was claimed to be capable of being programmed to deliver up to 27 kN of thrust and operated at frequencies from zero up to 2 kHz. A later design described by the same investigator (Brooks 1989) was claimed to be capable of controlling a payload of 1000 N over the frequency range from 1 Hz to 1 kHz. The maximum displacement claimed was 5mm with an acceleration of around 30 m s^{-2}. It was reported that static thrusts of up to 35 kN were measured.

Because reports of the performance of ER actuators are scarce and those devices that have been described have obviously been designed to meet widely differing specifications, it is difficult to place performance limits on ER actuators. Such limits are obviously essential so that actuators based upon ER fluids can be compared with actuators which use other forms of supporting energy. There seems little doubt that given the levels of thrust and bandwidths which have been reported, ER actuators offer a serious alternative to conventional forms of actuation. However, there are some factors which still inhibit more widespread commercial exploitation and these will now be summarized.

3.3.3 Factors Inhibiting Wider Commercial Exploitation

It is apparent from the foregoing discussion that it is feasible to construct linear actuators using ER fluid devices. The performance and relative simplicity of ER actuators seems to place them in a position to challenge hydraulic devices in many applications. However, the technology to produce

ER actuators has now been available for over thirty years and yet only prototype devices have been produced. This raises the question: what factors have prevented commercial exploitation of ER actuators?

There is no doubt that it is the nature of ER fluids themselves that is largely responsible for the delay in commercial exploitation. To help explain the reasons behind this statement, it is helpful to start by considering the basic requirements for a hydraulic fluid. As listed by Morse (1963), these are chemical stability, suitable viscosity and sufficient lubricity.

Early fluids used by Winslow (1947) consisted typically of paraffin contaminated with silica gel. Some water was added to provide electrical conductivity of between 10^6 and 10^8 $\Omega\,m^{-1}$. When judged against the criteria applied to hydraulic fluids, such suspensions are unsuitable in almost every respect. Reports suggested that the early fluids were highly unstable, liable to suffer rapid deterioration, and highly abrasive. These problems have been largely overcome by a shift to non-abrasive polymer particles in non-conducting silicone oils. However, there remain some features of ER fluids which prevent their straightforward application in some engineering applications and totally prohibit their application in others.

It has been stated (Stangroom 1983) that, as an approximate guide, the current drawn by a typical ER fluid will double for every 6 °C rise in temperature. This implies that there is a problem with thermal runaway which needs to be controlled in a practical device. The general problem of heating effects (electrical and mechanical) in ER fluids and temperature dependence of mechanical properties exists in all ER devices (Stevens, Sproston and Stanway 1988).

In some applications, the settling out of particles in the carrier fluid does not present a problem. For example, devices such as clutches and vibration dampers operating in the shear mode usually provide the necessary agitation to mix the ER fluid when relative motion is induced between the electrodes. In the flow mode, however, where there is no relative motion between the electrodes, settling out can represent a major problem. This obviously affects the ER valves which form part of an actuator arrangement.

The ratio of mechanical shear stress with and without an applied electric field is another major factor liable to reduce the effectiveness of control in ER devices. As long as the particles require a carrier liquid this will always represent a problem as the carrier liquid must possess a finite viscosity.

The long-term durability of ER devices and fluids is obviously of concern since there is no doubt that, by their very nature, ER suspensions will be more abrasive than conventional hydraulic and lubricating oils. Some studies have been carried out to assess the tribological properties of ER fluids (Lingard, Bullough and Shek 1989) but much work remains to be done.

Finally a factor which is liable to affect the performance of ER actuators operating at higher frequencies is the compressibility of the ER fluid. This factor was identified by Phillips (1969) and studied in detail using ‘delay-bond' modelling, an extension of the bond-graph technique. It was suggested by Brooks (1989) that compressibility of the ER fluid becomes significant at frequencies above 300 Hz and this may impose upper limits on attainable bandwidth. This is obviously an area where there is considerable scope for further work.

3.4 ER ACTUATORS—PRESENT DEVELOPMENTS AND FUTURE PROSPECTS

Not surprisingly, the main thrust of current development in the ER field is concerned with the fluids themselves.

Perhaps the most significant development is that due to Block (Jordan and Shaw 1989) who has developed a class of ER fluids in which the amount of water present is minimized. Since the presence of water can lead to dielectric breakdown, corrosion and high electrical power consumption, the development of nominally anhydrous fluids represents a major breakthrough.

To avoid settling out of particles, two approaches have been employed in recent years. One approach involves adjusting particle sizes using surfactants to inhibit particle interaction. The other approach involves matching the density of particles and carrier fluids. Matching suffers from the disadvantage that it can only be achieved perfectly at a single temperature due to the differing thermal expansion coefficients of the liquid and solid phases.

Finally, attempts have been made to overcome the problem of zero-field viscosity of fluids, mainly by employing carrier liquids of reduced viscosity. The problem with this approach appears to be that there exists a minimum carrier viscosity below which the ER effect ceases to exist. In some applications the problem of zero-field viscosity may have to be overcome by some other means, for example, by evacuating the space between the electrodes. However, note that measures such as this detract from the basic simplicity of the ER device.

If these problems can be overcome then there is no reason why electro-rheological actuators should not begin to challenge conventional hydraulic actuators in many engineering applications. For a discussion of ER and other forms of novel actuators, the interested reader is referred to the recent conference proceedings issued by the German VDI/VDE (1990) and the seminar organized by the IEE (1991).

3.5 CONCLUSIONS

Conventional hydraulic actuators offer a number of advantages over systems powered by alternative forms of supporting energy. However hydraulic systems generally require flow control valves which are complex and expensive components and contribute considerably to the overall system cost. One possibility for maintaining the performance of hydraulic actuators while reducing the cost is to employ electro-rheological fluids in place of the conventional hydraulic fluids.

The advantage of employing ER fluids arises from the fact that the flow control valves need involve no moving parts as flow is controlled by the application of an electric field. The performance available from ER actuators can approach that available from hydraulic units but there are drawbacks associated with ER fluids which must be overcome before more widespread commercial exploitation becomes feasible.

NOMENCLATURE

E	Electric field strength
H	Size of gap between flat plates
L	Length of plates
P_{O}	Pressure drop resulting from flow of Newtonian fluid
P_{ER}	Component of pressure drop due to ER effect
P_{TOT}	Total pressure drop
Q	Fluid flow rate
u	Fluid velocity
W	Width of plates
x	Flow direction coordinate
z	Transverse coordinate
η	Plastic viscosity
μ	Newtonian viscosity
τ_{y}	Yield stress
τ_{TOT}	Total shear stress.

REFERENCES

Block H and Kelly J P 1987 *US Patent Specification No 4687589*

Block H and Kelly J P 1988 Electro-rheology, *J. Phys. D: Appl. Phys.* **21** 1661–1667

Brooks D A 1982 Electro-rheological devices, *Chartered Mechanical Engineer* September 91–93

Brooks D A 1989 Devices using electro-rheological fluids, *Proc. 2nd Int. Conf. on Electro-rheological Fluids* (*Raleigh NC, August 1989*) eds J D Carlson, A F Sprecher and H Conrad pp 371–401

Bullough W A 1988 Integrated electron hydraulics, *Mech 88, Bicent. Congress and Expo* (Brisbane, Australia)
Burrows C R 1972 Fluid Power Servomechanisms (London: Van Nostrand Reinhold)
Duclos T G, Acker D N and Carlson J D 1988 Fluids that thicken electrically, *Machine Design* **January 21** 42–46
Eige J J and Fraser E C 1961 Feasibility study of vibration and shock exciter using electric field modulation of hydraulic power, *WADD Technical Report* 61–185
Institution of Electrical Engineers 1991 *Coll. Robot Actuators* (London, October 1991)
Jordan T C and Shaw M T 1989 Electrorheology, *IEEE Trans. on Electrical Insulation* **24** 849–879
Lingard S, Bullough W A and Shek W M 1989 Tribological performance of an electro-rheological fluid, *J. Phys. D: Appl. Phys.* **22** 1639–1645
Massey B S 1983 *Mechanics of Fluids* 5th edn (London: Van Nostrand Reinhold)
Morse A C 1963 *Hydraulic Servomechanisms* (New York: McGraw-Hill)
Phillips R W 1969 Engineering applications of fluids with a variable yield stress, *DEng Thesis* University of California at Berkeley
Stangroom J E 1983 Electrorheological fluids, *Phys. Technol.* **14** 290–296
Stanway R, Sproston J L and Stevens N G 1987 Non-linear modelling of an electro-rheological vibration damper, *J. Electrostatics* **20** 167–184
Stevens N G 1985 Electro-rheological fluids and their engineering applications, *PhD Thesis* University of Liverpool
Stevens N G, Sproston J L and Stanway R 1988 An experimental investigation of electro-rheological torque transmission, *ASME Trans. J Mechanisms* **110** 182–188
VDI/VDE 1990 *Proc. Actuator '90*, International Technology Transfer Congress (Bremen, June 1990)
Winslow W M 1947 Method and means for translating electrical impulses into mechanical force *US Patent Specification No 2417850*

4

Smart Power Devices

D H Paxman

4.1 INTRODUCTION

Users of semiconductor power switches to control actuators will have become aware that over the last decade, the semiconductor industry has been increasingly interested in the subject of 'Smart Power' devices which could influence their business significantly, either by replacing some electromechanical functions, or by simplifying the drive schemes for existing units.

It is the purpose of this chapter to give a very general introduction to the concepts of 'smart power', explain some of the technological aspects of the subject and give some illustrations of how the devices are constructed and designed. Smart power requires a much greater interaction between the circuit designer and the user and it is hoped that this description will help bridge the gap between the two partners.

Smart power is a term introduced in the 1980s to describe a family of new semiconductor elements which combine power devices with control components thus producing a power control system on a single silicon chip. The driving force of such endeavours is to open up new markets for semiconductor power, as well as offering the usual benefits of reliability, sophistication and cost effectiveness expected from an integrated circuit approach.

For convenience it is possible to classify the smart power family of devices into a set of members as illustrated in figure 1, based on the ratio of the power-to-logic (control) designed on the chip and the degree to which the system interacts with the external environment.

The simplest devices, protected switches, differ little from the conventional three-terminal discrete power devices. The amount of logic or control contained on the chip is very small, offering basic protection to prevent the

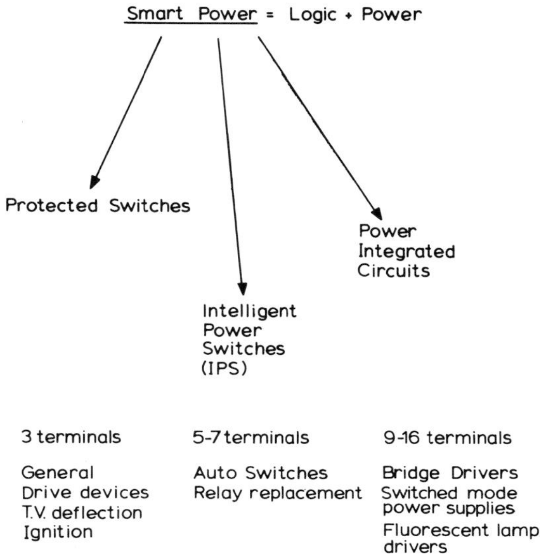

Figure 1. The smart power family.

user from inadvertently damaging the power element which would normally occupy about 80% of the chip area. They are usually used as an improved discrete device in a wide range of situations.

At the next level, the intelligent power switch (IPS), the degree of control has risen and the device not only contains protective functions but has the ability to signal information about the status of the unit back to a central control system. The device usually retains a power device appearance but can be distinguished by having an increase in the number of connection pins to deal with the extra logic input/output [5–7]. Such switches would normally have a ratio of logic-to-power of roughly 40%–60%. Devices of this type are of particular interest as relay replacements or driver units for a range of loads, particularly in the automotive industry.

Finally, the power integrated circuits (PICs), as the name implies, resemble ICs in outline having considerable interaction with both control and loads to offer sophisticated power handling functions to the users. The device may contain a power device sufficient to drive the load directly or may be used with extra discrete components, as a bridge driver unit for example. Devices of this type are usually found as ballast systems for fluorescent lamps or complex power supplies.

All these devices have become possible technological developments, often derived from integrated circuit technologies which have been incorporated into power device manufacturing processes over the last decade. Before examining the technological aspects of the devices it is advantageous to give an outline of the functions required since this gives a measure of the complexity required in the silicon.

The types of protective circuits demanded in such switches usually detect malfunctions such as excessive temperature, currents and voltages in which case the unit is turned off safely. Other circuits may help stabilize the device by, for example, limiting the maximum current which may be drawn. Other non-destructive malfunctions, for example open loads, may be detected and an indication given. The exact condition of the device may, if required, be signalled via an extra status pin using a simple logic system to deduce the fault.

4.2 COMPONENT REQUIREMENTS FOR SMART SWITCHES

The components required for these circuits have to be created *in situ* in the processing of the silicon. The task of the semiconductor engineer is to optimize the design of the devices at the same time as minimizing the costs.

The types of component called for in power applications can be categorized as follows.

(1) Power devices—suitable for all load requirements.
(2) Logic devices—low voltage for compact circuits.
(3) Interconnection devices—high-voltage MOS or bipolar devices which can connect the power device control circuits to the logic.
(4) Zener diodes—voltage setting.
(5) Bipolar devices—temperature measurement
(6) Resistors and capacitors.
(7) Isolation structures—these are regions which can separate the high- and low-voltage parts of the circuits.

These devices are described in more detail in the following section.

4.2.1 Power Devices

The power handling component is the major element within the smart switch and its design exercises a dominant influence on the overall technology chosen.

The most widely used power device is the power MOSFET or some variant of this basic switch. The growth of smart power as a practical solution to switching has depended very much on the introduction of this new type of power device in the 1980s.

Figure 2(*a*) summarizes the main features of the power MOS. The device consists of an array of cells (square, octagonal or circular) at a density of up

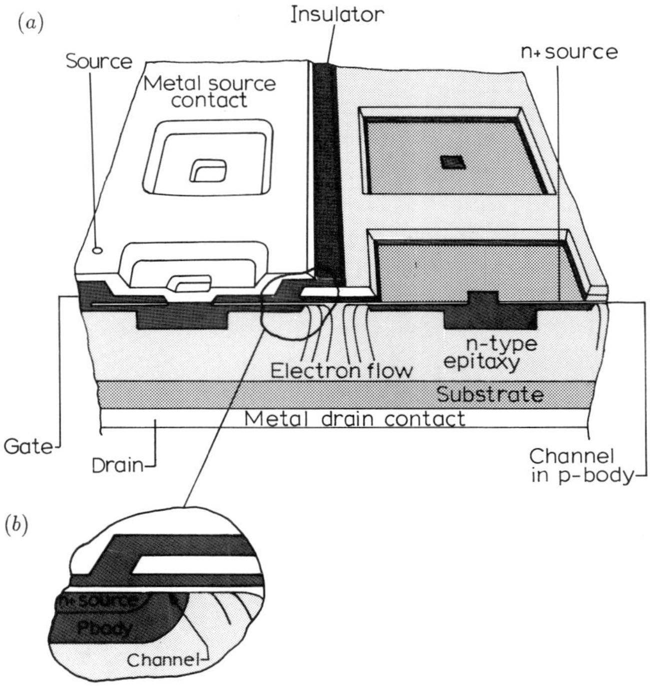

Figure 2. (*a*) A summary of the main features of the power MOS. (*b*) Simplified three-dimensional cross-section of an *n*-channel enhancement POWERMOS transistor.

to 5000 mm^{-2}, with each cell containing a small MOS device as shown in figure 2(*b*). Each cell consists of a central source region from which the current flows laterally under the controlling gate electrode (made of polysilicon on an oxide 500 Å thick) into a lightly doped epitaxial layer in which the current turns through 90° to flow vertically to the drain contact formed by the heavily doped substrate on the back of the slice.

Such power MOS devices are available in a wide range ofvoltages (50 V to 1000 V) determined by the thickness of the voltage supporting epitaxial layer. The device current is controlled by the gate electrode and the modern power MOS is optimized so that the control can be exercised by logic level signals (1–5 V < 10 mA). The device characteristic is essentially resistive and a typical 50 V device will have an on resistance of 300 mΩ mm^2 so that 10 mm^2 of cells are required for a 30 mΩ on resistance.

While this vertically conducting device uses the surface area very effectively, the fact that the substrate plays an active role in the device operation can cause difficulties in situations where the circuit calls for isolation of the various power control elements. The alternative lateral version (lateral DMOS) in which the source and drain are on the top surface is shown in figure 3. Such devices are more difficult to design for low on resistance at low voltages but compete effectively with their vertical counterparts above 400 V.

The combination of low drive requirements, reasonable power handling capability and constructional features which contain MOS elements has made these devices the core of many smart power technologies.

It should be mentioned at this stage that there are a number of variants of the MOS power device in which bipolar injection takes place so that the device can offer the combination of ease of MOS control with the very low resistance of the bipolar devices. As discrete versions of these devices mature, it is to be expected that they will gradually appear in smart power formats.

Finally, there are units built around the more conventional bipolar transistors. These tend to be confined to the protected switch type of device since the basic construction of the bipolar device does not lend itself so easily to the construction of logic devices.

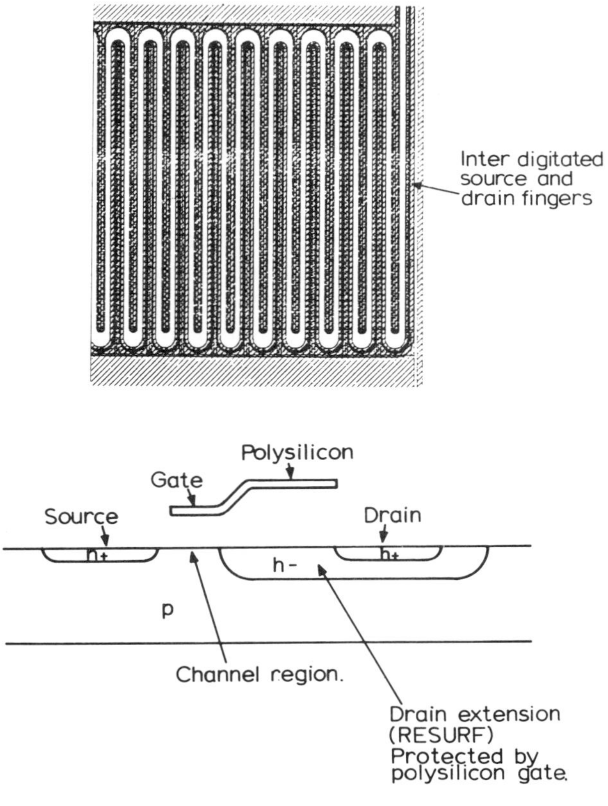

Figure 3. Lateral power MOS design.

A summary of the basic constructional features of the power MOS shown in figure 2 can be given as follows:

Gate	polysilicon (doped for high conductivity) on 500 Å oxide.
Source	n^+ region diffused below the polysilicon edge (auto-aligned).
Channel	(p body) p^+ diffusion deeper than and surrounding the n^+ source.
Shorting system	the n^+ source, the p body and the epilayer form a parasitic bipolar transistor. To prevent this device from turning on, the centre of each cell contains a region of high p^+ doping connected to the p body (base) and an n^+ contact to the source (emitter). The metallization is arranged to short these together.
Typical dimensions	source window: 10–20 microns wide. polysilicon track: 3–8 microns wide.
Junctions	0.8–3.5 microns.
Mask stages	5–7 masks.

4.2.2 Logic Devices

Low-voltage logic The essential feature of the low voltage devices is the small size, with very easy on/off control by means of an MOS gate. The ability to provide both n- and p-type MOS devices is important for the provision of complementary switching functions.

The logic devices used in many smart power products are versions of the standard complementary MOS types used in many integrated circuits adapted to the technology as described by Milne [1]. Figure 4 illustrates the *IV* characteristics of the basic devices. Typical dimensions of the channel region below the gate would be 1–5 microns long (source to drain) and 1–10 microns wide. In smart power applications the designer is mainly concerned with the threshold voltage (voltage to turn the device on) and the flatness of the saturated characteristics. In many circuits, use is made of the constant value of I_{ds} for a fixed V_{gs} as the V_{ds} varies.

With such small devices the maximum voltage which can be supported between the source and the drain (breakdown voltage) is limited to 5–20 V and these devices would be insufficient to deal with any high voltages in circuits built around the power device. Increasing the distance from source to drain increases the resistance of the device and does not materially alter

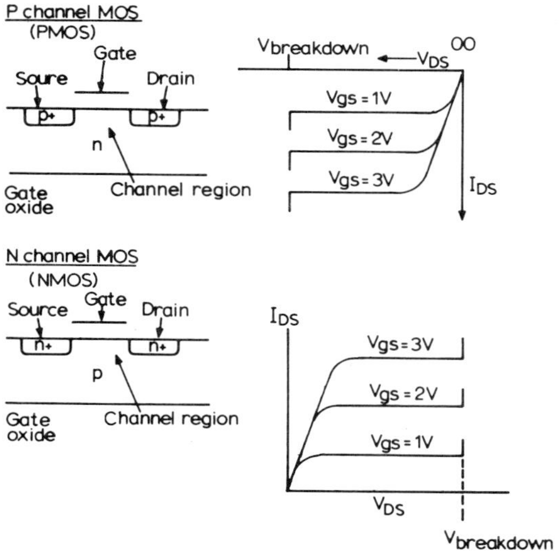

Figure 4. Simple MOS logic devices (low voltage).

the voltage available. In order to combine good channel properties with special high voltage, special high-voltage logic devices are required.

High-voltage logic High-voltage CMOS devices are constructed by incorporation of an additional lightly doped region (drain extension) between the source and the drain, as shown in figure 5. This region is required to provide a volume of silicon in which the electric field can be distributed in a controlled manner to support the applied voltage without exceeding the breakdown field of 2×10^7 V m^{-1} for silicon. RESURF is the term describing these layers designed to REduce the SURFace electric fields but keep the bulk electric field high so the voltage is supported in the minimum volume. The lightly doped drain regions require considerable skill in design and fabrication to achieve the correct combination of properties [2].

The characteristics of these devices are exactly the same as for the low-voltage versions with the breakdown voltage lying between 50 and 500 volts depending on dimensions and other physical properties. Often, the drain region of these devices is quite a complex structure to cater for the difficult conditions experienced.

Both the high- and low-voltage CMOS logic devices have bipolar equivalents replacing the gate electrode with a base contact to the region between the source (emitter) and drain (collector) and some schemes of logic use these devices for control functions.

(a)

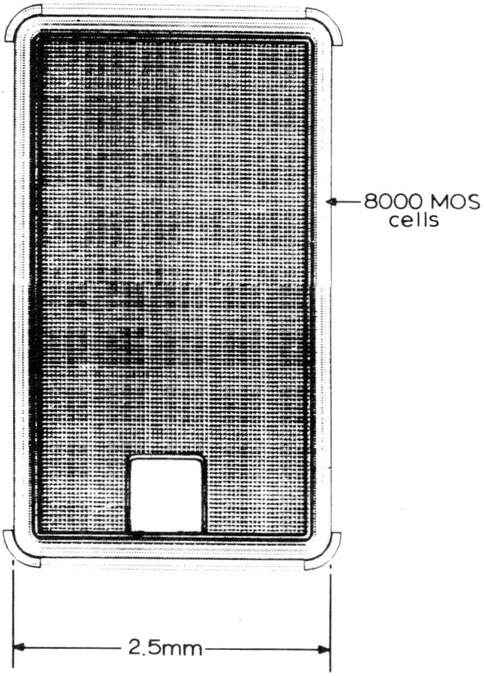

(b)

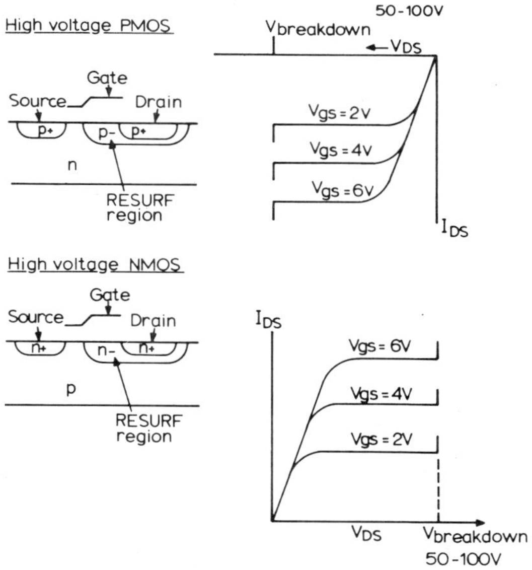

Figure 5. (*a*) Plan view of power MOS. (*b*) High-voltage PMOS and NMOS.

4.2.3 Other Bipolar Components

For most circuits, other devices are required for rectification, protection and temperature sensing based on bipolar designs. The most common components are shown in figures 6(*a*) and (*b*). Bipolar transistors with gains in the region 100–300 can be made.

If highly doped p^+ and n^+ layers are used in the diodes, then a kind of Zener device with a controlled breakdown in the region of 5 to 15 V can be constructed for voltage reference tasks. Both the base/emitter and base/collector diodes of the bipolar transistors can be used if the configuration is suitable.

4.2.4 Other Components

Devices like capacitors and resistors can be built from basic combinations of p^+, n^+, polysilicon and oxide layers in a variety of ways.

Capacitors are shown in figure 7 with polysilicon and aluminium electrodes and gate or other oxides for the dielectric. The gate oxide is most useful for a high capacitance per unit area (7×10^{-6} F cm^{-2}) for 500 Å oxides), but the applied voltage should not exceed the normal safe limit on gates of 15 V.

Resistors can usually be made from diffused areas utilizing the most lightly doped regions for highest resistance values. Care has to be taken with such resistors since parasitic diodes are easily formed down to the substrate. For full isolation, polysilicon has considerable advantage but the resistance range is limited if doped polysilicon is used.

It will be observed that in this range of devices the construction has repetitively used certain basic units: diffusions, oxides, polysilicon, etc. By

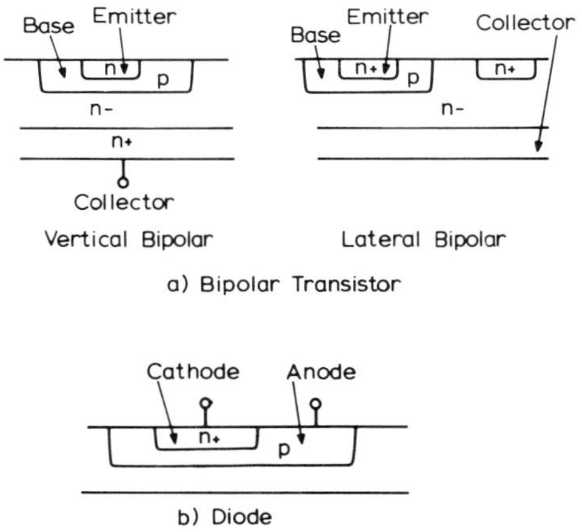

Figure 6. Bipolar components.

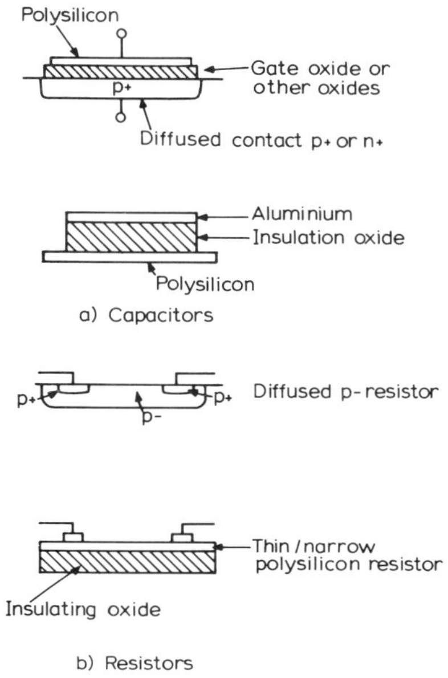

Figure 7. Assorted components.

Table 1.

Layer	Component
n^+	Source of power MOS Source and drain of all NMOS Emitter/collector of lateral bipolar (npn) Cathode of diodes, Zeners
p^+/p^{++}	p body of power MOS Source and drain of all PMOS Emitter/collector of lateral bipolar (pnp) Anode of diodes, Zeners
Polysilicon	All gates Resistors
Gate oxide	All gates Capacitors
p^- RESURF	Drain extension of high-voltage PMOS p^- resistors (approximately)
n^- RESURF	Drain extension of high-voltage NMOS n^- resistors (approximately)

using these basic blocks skilfully it is possible to keep the number of component layers to a minimum, thus ensuring reasonable cost but inevitably losing some of the degree of freedom offered by a full IC technology.

Table 1 gives an indication of a typical set of layers and their use in a simple power MOS-based smart switch.

4.3 ISOLATION TECHNOLOGIES

Because of the different electrical conditions within the power and logic parts of smart power devices, an ideal design is to arrange complete electrical isolation between the two sets of components, as shown in figure 8. This gives the designer complete freedom of choice with the circuits and logic components as well as enabling the fullest optimization of the power elements. This is particularly true for cases where vertical power devices are used, since the substrate (back of the slice) is either connected to the load or the power supply and is electrically active.

This isolation can be achieved physically by means of dielectrics, or an equivalent electrical isolation can be created wholly or partially by using the properties of electrical junctions. Dealing with these in turn, we have the following.

Dielectric isolation Dielectric isolation with a complete surrounding of the active logic by an insulator from the power part of the chip can be obtained in a variety of ways, and considerable effort is expended in devising cost-effective solutions. One of the most successful methods is illustrated in figure 9(a(1)) in which single-crystal regions are obtained by first selectively etching single-crystal silicon into raised regions. These are oxidized to cover with SiO_2, refilled with polysilicon as a supporting layer and finally back-polished, leaving isolated regions which, if inverted, provide isolated single-crystal tubs in which a wide variety of MOS, bipolar and power

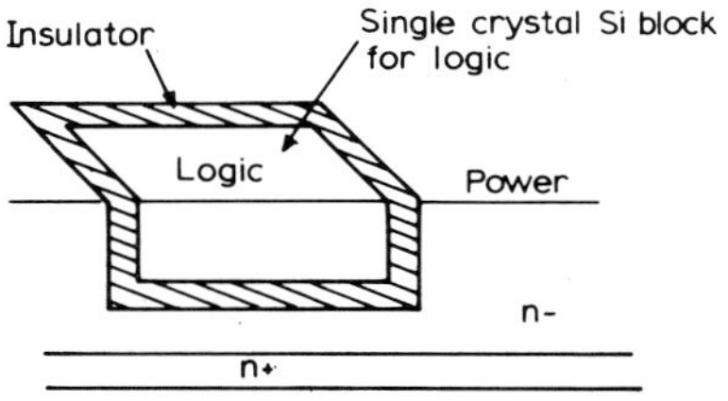

Figure 8. Ideal isolation scheme.

devices can be made [3]. As drawn, it would seem difficult to obtain a vertical conduction device within the technology, but if at stage (*a*) the surface is doped with an n^+ layer, the resulting tub has a lining of high conductivity which can be used for back contacted drains or collectors as shown in figure 9(*a*(2)).

A similar result can be obtained by starting with a 'sandwich' system of a supporting substrate, SiO_2 insulator and an active epilayer on the surface which can be obtained by physically attaching the layers together as shown in figure 9(*b*). With the active epilayer supported on a substrate, a strong bond to an SiO_2 layer can be achieved if the mating surfaces are polished flat and are microscopically clean. Polishing or etching is required to remove the upper supporting layer to reveal the working layer. The technique requires further trenches to be etched down to the insulator if isolated 'islands' are required for the various components. As in the previous case an n^+ layer can be inserted to produce the back contact if required [4]. Both techniques are used in current products. In the research phase, alternatives are being investigated. Probably the most likely to be used in the near future is a method based on the recrystallization of a polysilicon layer formed on top of an insulating layer which in turn is on top of an active device layer. As shown in figure 9(*c*), the hole in the dielectric is to ensure the correct crystallization of the polysilicon into a single crystal in which MOS devices can be made. The advantage of the method is that it is basically cheap and enables the starting layer to be used as conventional material for the power part of the device [5].

Junction isolation Electrical isolation can be obtained to quite high voltage levels by surrounding logic parts of the devices with screening regions of opposite polarity. For example, as shown in figure 10, a p region (p well) can be isolated by surrounding it with an n^+ buried layer completed to the surface with deep n^+ diffusions. The buried layer can be obtained by implantation at an appropriate stage during the growth of the epilayer. Such techniques can be extended to make a variety of p wells and n wells and retain the ability to leave regions free for the construction of vertically conducting devices. Considerable care has to be taken with electrical contacts to these additional layers since parasitic devices like thyristors are easily formed in the multilayer structures. (A thyristor consists of four semiconductor regions in a p/n/p/n arrangement forming a coupled pair of p/n/p and n/p/n transistors.) This arrangement provides mutual feedback, so the system can be brought into a heavily conducting on state in which external control is impossible [6]. Once in this condition only removal of the power supply will turn the system off. The high currents and rapid transients experienced in the power devices switching must not trigger such devices into a 'latched up' condition in which control is lost. Though complex, these methods are well known from integrated circuit technologies

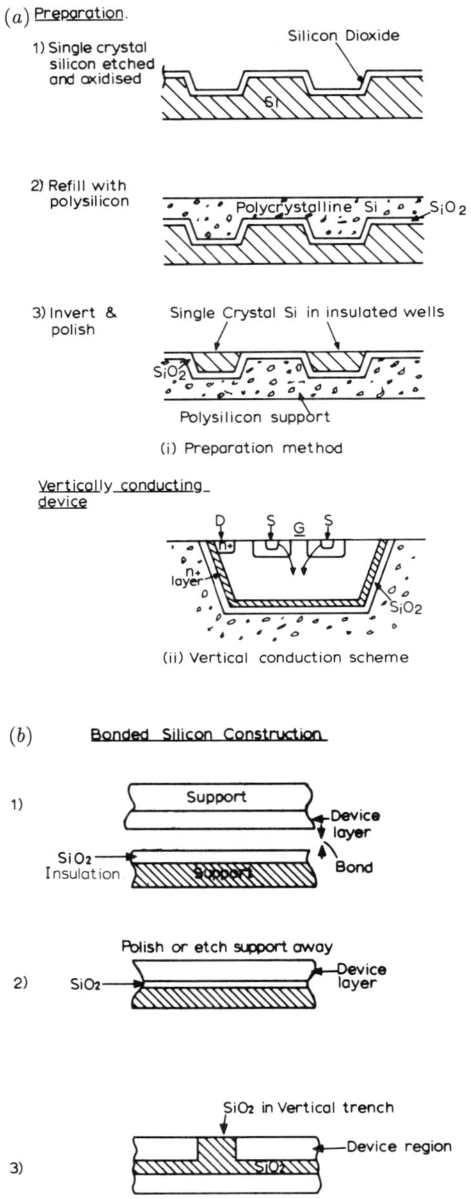

Figure 9. (*a*) Dielectric isolation. (*b*) Dielectric insulation (bonded silicon). (*c*) Dielectric isolation (recrystallized silicon).

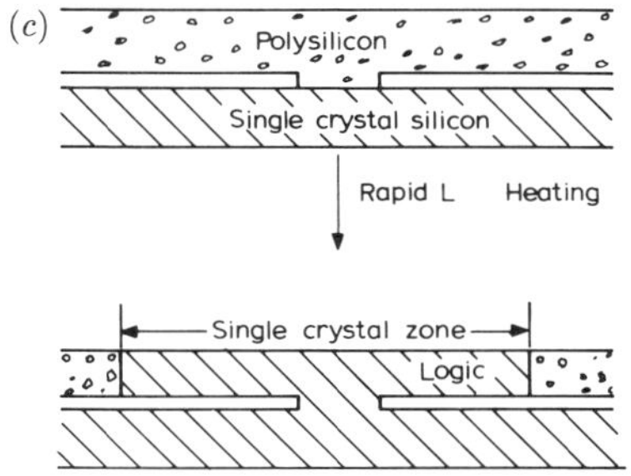

Figure 9. *Continued.*

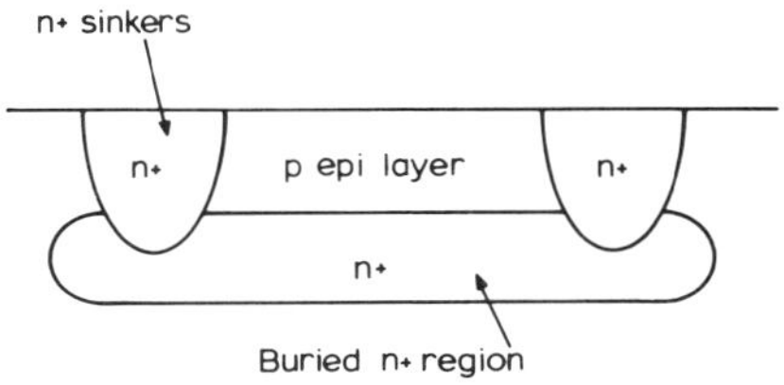

Figure 10. Junction isolation.

and advantage can be taken of the techniques provided the cost can be offset by the sophistication of the product.

Self-isolation This is a technique which avoids the technical complexity of the junction isolation while retaining its electrical behaviour. The circuit designer has no longer complete freedom of polarities and voltages within the design but usually sufficient flexibility is retained to allow useful systems to be constructed. The isolation regions are obtained by diffusions or their need is avoided by the use of suitable choice of power supply voltages.

Figure 11 shows a simple scheme. The NMOS device is formed within a diffused p region which is designed to allow the correct operation of the NMOS even if large reverse bias is applied to the drain with respect to the active substrate.

The complementary PMOS device is operated within the basic n^- epilayer which is electrically possible provided the source and drain potentials never exceed the supply voltage and have a reverse voltage not greater than the breakdown of the $p^+/n^-/n^+$ diodes. Thus it is usual to operate the logic supply referred to the external supply to minimize reverse voltage effects in high side switches. By extending the idea of RESURF it is possible to construct the devices in wells which will enable the full high voltage logic to be fabricated.

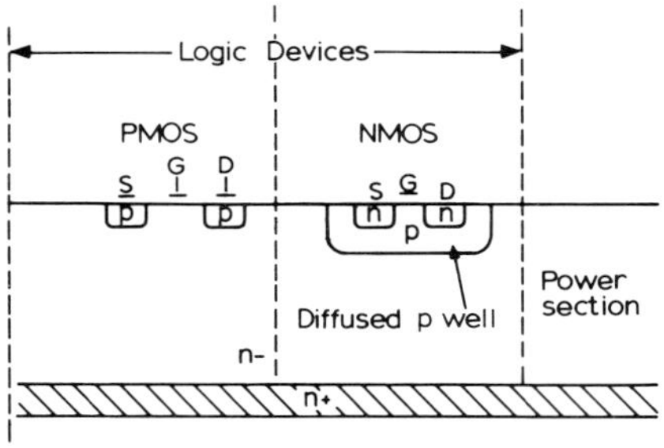

Figure 11. Self-isolation technique.

Table 2.

Technology	Use	Comment
Dielectric	High-voltage circuits demanding high isolation	Complex starting material
Junction isolation	Complex IC-type logic	Expensive but well understood
Self-isolation	Simple logic with circuit restrictions	Cost effective but limited intelligence

The technology has the very large advantage that the structures are obtained entirely by diffusion and with ingenuity can be produced with the minimum of masks [7].

Table 2 summarizes the position of the various methods.

4.4 CIRCUIT ASPECTS

Circuits can be divided into power control functions and protective/signal functions.

4.4.1 Power Control Functions

These deal with the turning on and off of the power device and usually comprise an input circuit and a power device driving system.

Input circuit Normally the input circuit is designed to match with the output of a control system based on standard TTL logic. Typical input circuits will operate at <5 V and <1 mA drive current referred to ground. To ensure correct operation, hysteresis is built into the circuit so the turn off voltage is less than the turn on. In order to reduce quiescent current the input circuit will often 'wake up' the internal functions, in particular, power supplies.

Power device drive The configuration and type of power device determines the functions required. For MOS power devices the gate drive can be as low as 5 V in which case, for a low side switch, the power device can be directly turned on by the input. For the more commonly used high side switch the system has to be more elaborate since the gate voltage has to be >5 V above the source to turn the power device on. This requires the generation of an extra voltage on the chip greater than the power supply. A typical system is shown in figure 12. An oscillator running at 50 to 100 kHz is used to operate a charge pumping system in which a small capacitor is alternately charged by the power supply and discharged via a rectifier into the storage capacitor formed by the gate oxide of the power MOS [8]. This will generate an extra voltage close to the logic supply voltage on the gate. More elaborate double and triple chargepumping schemes can be used to increase the voltage. By inserting time constants within the gate charging and discharging systems the rise and fall of the power device current can be controlled. The drive system also contains high-voltage lateral logic devices for rapid switch off to the power device in emergencies.

4.4.2 Protective Functions

Most smart devices will contain a selection of safety features from the following types.

Overtemperature The difficulty with temperature measurement in smart power devices is the variability of all the components with temperature. The

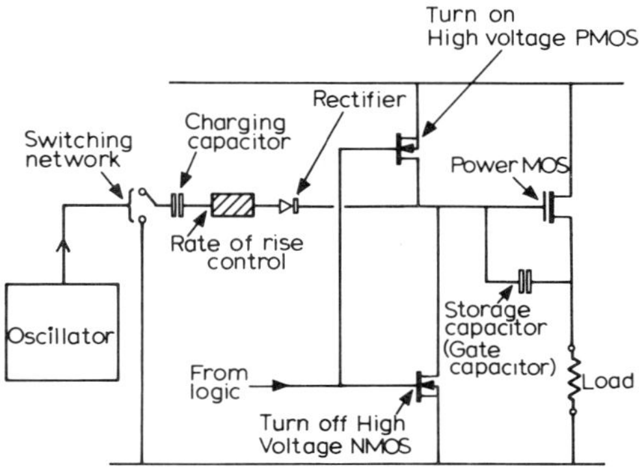

Figure 12. Power device drive system.

problem is in obtaining a stable reference independent of temperature against which the temperature-varying quantity can be compared in order to determine a triggering point. In very simple (and necessarily inaccurate) schemes, internal leakage current of a diode can be used since this rises very rapidly with temperature above 150 °C. As in figure 13(a(i)), this current passed through a resistor can be used to trigger an MOS gate when the threshold voltage is reached. Since both the resistor and threshold values for the MOS are variables it would be necessary to adjust R (by laser trimming) if accuracy was required.

Similarly, the leakage current in a thyristor can cause turn on at a particular temperature and thus can be used to turn off a device [9]. If more accuracy is required, particularly if temperature hysteresis is to be built-in, then considerably more complex methods are required. These are normally based upon variants of band gap reference systems well known in IC technology (see, for example, [10] for a full discussion). The essential features of a scheme based on these ideas is shown in figure 13(a(ii)). The system relies on the properties of the two diodes D_1 and D_2 which are considerably different in area. For such a diode the forward voltage drop is given by

$$V_{\mathrm{Diode}} = \frac{kT}{q} \ln\left(\frac{J_{\mathrm{D}}}{AT^{\gamma}}\right)$$

where J_{D} is the current density.

In the scheme shown, a current mirror system is used to force the same current through the two diodes in which case (for areas in ratio 1:10)

$$\Delta V = V_{\mathrm{D}_1} - V_{\mathrm{D}_2} = \frac{kT}{q} \ln\left(\frac{J_{\mathrm{D}_1}}{J_{\mathrm{D}_2}}\right) = \frac{kT}{q} \ln(10)$$

Thus the output ΔV is proportional to T and can be used for temperature sensing. Accurate temperature sensing requires considerable design skill and care in laying out components for getting matched operation.

Over/under current protection In MOS devices, a convenient method of measuring current is to use a sense cell built into the main power device. A single or small group of cells may be isolated as a 'sense cell' by defining the surface metallization. This sense cell will have the same gate voltage as the main part of the device. A comparator can be connected to the sources of the main unit and the sense cell to ensure that the source voltage on each is the same. Under these conditions, the current drawn from the comparator will be in the ratio of the numbers of cells in the main and sense devices. The resulting scaled current is measured by passing it into a resistor to create a voltage which can be compared with a set value to determine the trip point

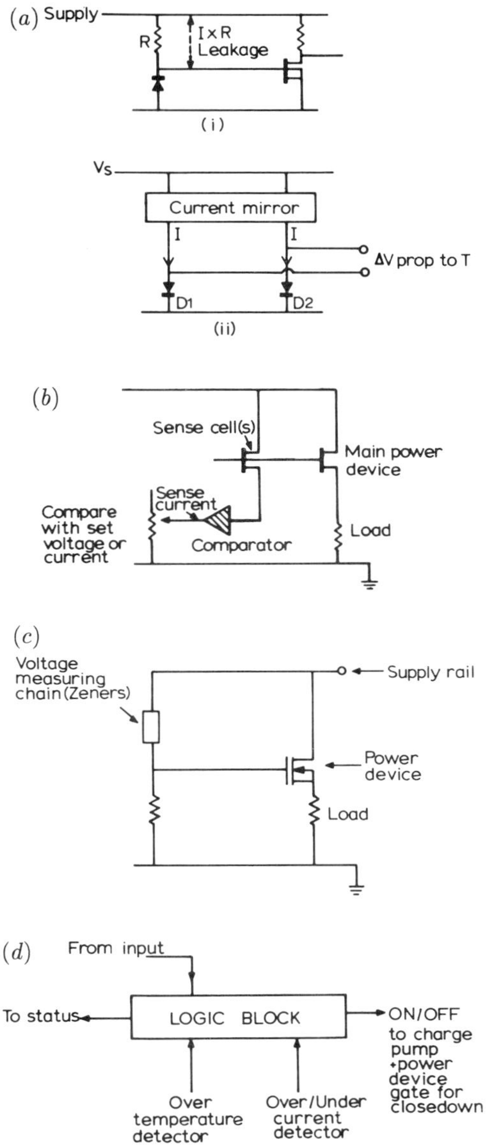

Figure 13. (*a*) Over-temperature protection. (*b*) Over/under current protection. (*c*) Over-voltage protection. (*d*) Logic scheme.

(figure13(*b*)). Such a scheme can be used for both high and low currents. The greatest difficulty in such protective schemes is in setting the limits in a practical system which are accurate enough to distinguish safe and unsafe operating regions for the switch and load. For example, in many lamp switching operations the inrush current into a cold lamp may well resemble a short circuit. Often a delay time is built into the detection loop to allow time for true lamps to warm up and reach the correct operating current before the decision point is reached. Considerable information on the fault conditions is needed to be able to set up practical schemes using electrical data in this way.

Overvoltage protection In the case of a poorly regulated supply subjected to interference, the rail voltages may be subjected to voltage 'spikes' containing limited energy but exceeding the breakdown voltage of the power/logic devices. To prevent damage, Zener voltage clamping devices can be incorporated into the system, but a cheaper alternative is to use the power device itself as an energy absorbing structure. As shown in figure 13(*c*), a chain of Zener devices connected to the gate of the power switch can detect excessive voltage and turn the power device on for a brief period to 'dump' the energy spike into the load.

Reverse voltage protection In many battery-based systems, provision can be made for the units to survive this condition (possibly for a limited period). For dielectrically isolated devices there is little difficulty, but using the simple self-isolation technologies, the p well structures have to be prevented from forward injection, at least to excessive currents. In many cases a simple limiting resistor in the p well supply suffices, though there are considerable practical difficulties with such schemes.

4.4.3 Control Logic

Each of the safety circuits provides an error signal which is combined within a logic circuit designed to give priority to the particular error messages, and to shut down the power device via the drive system if necessary. In addition to turning off the power, the unit also activates the status output. Thus, as shown in figure 13(*d*), the logic will ensure close down for the case of short circuits or overtemperature signals, but merely indicate a fault for open circuits.

4.4.4 Status Indication

The status pin on the device is usually connected to ground in case of error, and in simple schemes sufficient current capability would be provided to operate an LED. More information can be obtained from the status output if the signal is examined in conjunction with the state of the input signal. For example, both overtemperature and short circuit will turn the unit off, and indicate an error on the status pin. To distinguish between the two cases the system can be arranged so that for a temperature fault the error signal will

Table 3.

Mode	Input	Status	Output
Normal	L	H	L
	H	H	H
Shorted load	L	H	L
	H	L	L
Thermal shutdown	L	L	L
	H	L	L
Open load	L	L	H
	H	H	L

H = high output.
L = low output.

not be altered if the input is turned off and on, whereas for the short circuit, the error signal will change in synchronization as the input is switched. A wider range of fault indications can be obtained, using in-built time delays to separate the various faults. In most cases the manufacturer offers a simple 'truth table' which allows the various faults to be diagnosed as shown in Table 3.

4.5 PRODUCT REVIEW

Inevitably, any detailed Product Review dates rapidly as the market develops. However, it is possible to distinguish between types of products. Most major semiconductor manufacturers are involved in smart power designs. Often the catalogue products are the result of types which have been developed in conjunction with specific customers which are then accepted for the more general market. Since the devices can to some extent be modified to suit particular specifications it is possible to consider specific designs for large scale orders of a particular modification.

It should also be noted that many current units may initially be in a hybrid form with a number of internally connected chips, since structures are precursors to the full integration on a single chip.

4.5.1 Protected Switches

A number of manufacturers have described variations of power MOS [11] and bipolar [12] power devices which contain additional protection. Most protected types rely on simple drive protection (gate protection for MOS) and very simple temperature control by resistors and bipolar devices to limit currents as temperature rises, or even use a temperature-triggered thyristor to turn the device off at elevated temperature.

The interest in these devices is widespread but the low level of control and difficulty of setting safety limits are a problem.

Figure 14 gives a block diagram of a typical protected switch. Protection of the input circuit would be a control on the drive system to prevent excessive currents or voltages developing. The load/power device would be protected by having some form of current limiting system to ensure the device remained within safe operating regions and/or some form of temperature measuring system to prevent external short circuits from damaging the semiconductor. These circuits operate automatically to ensure that provided the device is operated within specified limits, the power element will be indestructible. The total number of components in such a switch would be between five and ten, so the circuit has to be particularly simple.

There is no form of status indication since the device is a simple three-terminal structure and the power for the internal logic has to be drawn from the input pin.

4.5.2 Low Side Switches

Low side switches form an important class of devices useful for driving a wide range of inductive and resistive loads. In the low-current range, devices are effectively specialized versions of ICs often having one or more

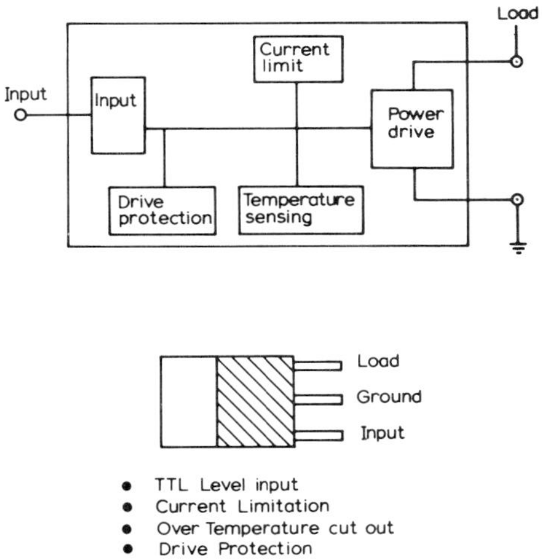

Figure 14. Protected switch.

outputs. As the current rises, the devices become very similar in outline to the high side switches with similar protective functions. For vertically conducting devices the slice back (drain or collector) is connected to the load. These devices are technologically demanding to manufacture in simple self-isolated schemes [13].

As opposed to high side switches a separate supply voltage (sometimes derived from the controlling IC system) will be needed. Figure 15 shows a block diagram of a typical low side switch.

For simple motor control circuits can be included for speed control by pulse width modulation of the power device with extra pins for external speed setting. Typical low side switch specifications are given in table 4.

4.5.3 High Side Switches

These are often referred to as Intelligent Power Switches (IPS). A wide range of these devices is available since they have been extensively developed for use in the automotive industry to drive lamps, relays, wiper motors, etc.

Figure 16 shows a block diagram and photograph of a typical unit that contains considerably more circuitry than a low side switch.

The power control part is the chain formed by the input, drive control and the power device. The drive control may include elements for current limitation and slow rise and fall circuits to tailor the power device output to the demands of the load.

The protective functions comprise detection schemes for short circuits, open circuits, and excessive voltage all based on internal measurements made around the power device. Temperature is measured independently. These circuits will close down the power device in the event of a malfunction. The purpose of the logic system is to ensure that the correct sequence

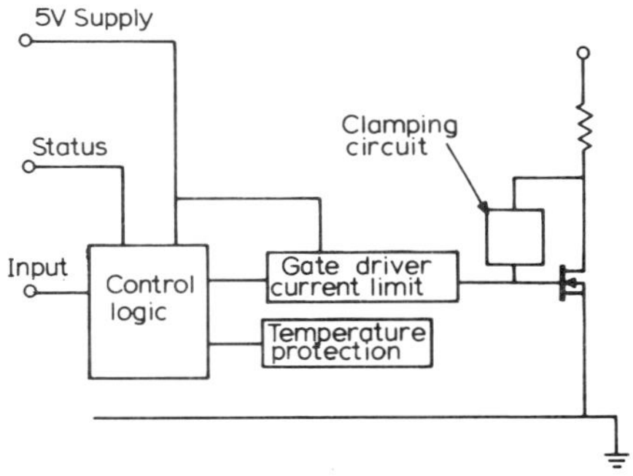

Figure 15. Low-side switch.

Table 4.

Package	Modified TO220
Output power device	Power MOS open drain NPN open collector
Current range	2–10 A
Saturation voltage (NPN)	0.8
On resistance (MOS)	50–200 mΩ
Supply voltage	5–45 V
Output voltage (Power device)	50–70 V
Output current	1–4 A
Protection	
Overvoltage Undervoltage	} power deactivated
Overcurrent	Current limitation
Shortcircuit	Shutdown
Thermal excess	Shutdown
Status	Status indicated by status pin low and simple truth table

of the control takes place and to provide an output to the status pin. The status pin condition in conjunction with the situation of the input pin can be used with a simple truth table to allow the user to determine the cause of the device shut down. An onboard power supply is present so the logic devices can be operated at low voltage compared with the voltages present on the power device to enable small compact logic to be used. In some cases allowance has to be made for the possibility of reversed supply voltage, and appropriate protection provided. In the case of a poor quality supply rail producing overvoltage ‘spikes’ it may be possible to turn the power device on briefly to absorb excessive energy to protect the logic circuits. The total number of components in such a typical IPS would be between 200 and 400. Typical high side switch specifications are given in table 5.

Technologies in such switches range from the simple self-isolated diffused technique through to the more complex buried layers, or dielectric isolation methods depending on the manufacturer’s preference. Most control circuits are in some modified CMOS form, though some technologies allow mixtures of bipolar and MOS components. For most applications, the output device is a 60 V power MOS.

Units of this type can be used in a wide variety of non-automobile applications wherever a high side power switch is required, for example in driving solenoids, printer hammers, needle valves and relays.

For descriptions of particular high side switch schemes see references [14] to [17] and the manufacturers’ literature.

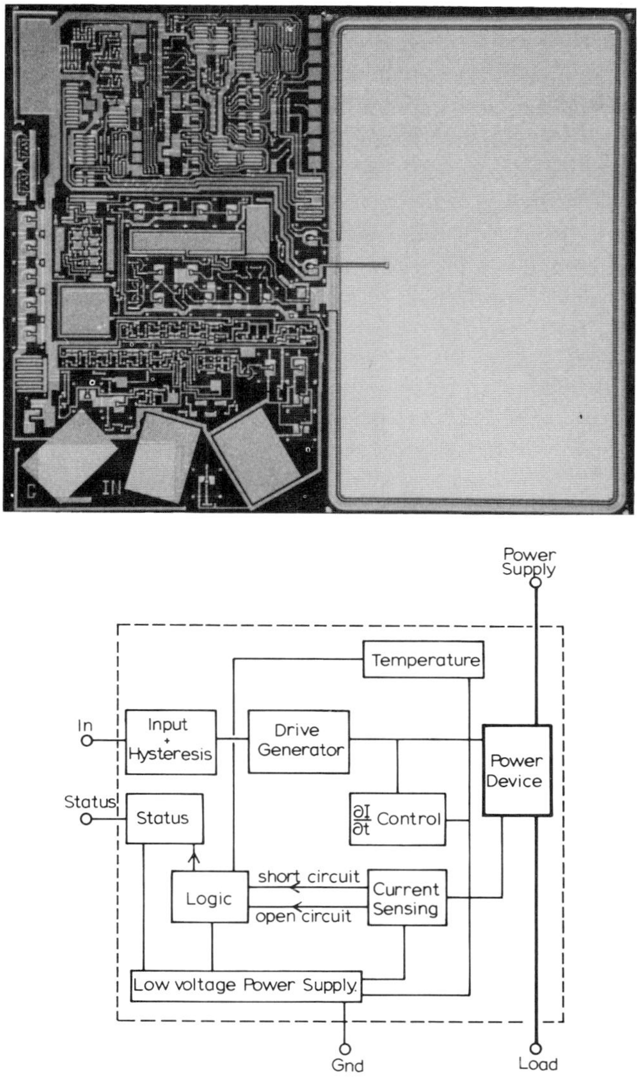

Figure 16. Intelligent power switch (IPS)

4.5.4 Bridge Units

The four-switch bridge unit (H Bridge) shown in figure 17(*a*), is an important component for many power control applications. In principle, such a unit can be built up from combinations of high and low side switches, but it is extremely difficult to manufacture in fully integrated form as access has to be obtained to the four separate collectors/emitters,

Table 5.

Package	Modified TO220 5–7 leads
Power device	Vertical MOS
Operating voltage range	5–18 V
Current rating	1–10 A
Max. voltage obtainable	45–70 V
Surge current	30 A
Control voltage	TTL with hysteresis
	V_{on} >3.0 V
	V_{off} <2.5 V
Protection circuits	
Temperature limit	160 °C
Hysteresis	30 °C
Overvoltage protection	>40 V
Short circuit protected	
Reverse supply protected	
Status	Status pin low for fault
	Capable of driving ~1 mA
Diagnostics	via a truth table sensing the states of the input and status pin to derive the fault information

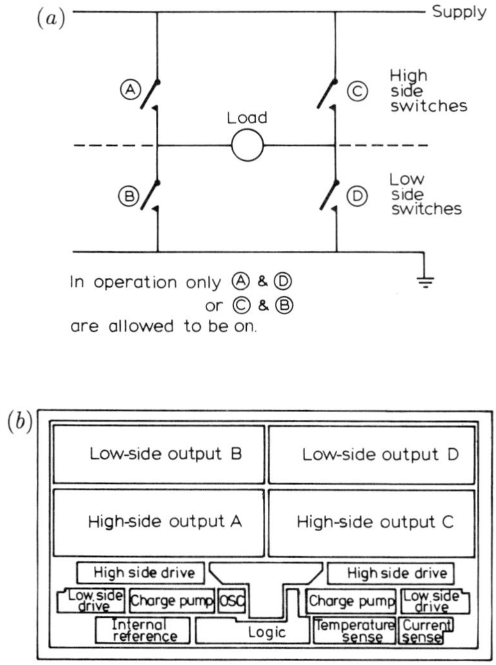

Figure 17. (*a*) Simple H-bridge. (*b*) Monolithic H-bridge.

drains/sources of the power devices. At this stage in development bridge units are often made fróm hybrid assemblies with separate chips for the power devices in the low side switches. Alternatively, the unit can be built from a bridge driver with four separate individual power switches.

Fully integrated units require lateral devices for the output stages, or at least technologies with buried layers or dielectric isolation to allow access to the individual power elements. Because of this, most monolithic full bridge systems are confined to low current ranges since relatively high-resistance lateral DMOS power devices are used. Alternatively, modified lateral DMOS devices in which some bipolar injection is introduced (LIGTs) can be used since these have an improved on resistance [18]. The packages are necessarily more complex than for most units described so far, and specialized packages having 10–14 pins have been developed by most manufacturers.

As shown in figure 17(*b*) an ideal monolithic bridge has four switches which are operated in sequence for simple reversal of a DC motor, or at high frequencies if the load is part of a power control circuit (switched mode power supplies) or for a fluorescent lamp. It is extremely important, particularly at high frequencies, to ensure that the switches on one half of the bridge, A and B or C and D, are never used simultaneously, so such power control systems require considerable monitoring and supervisory electronics making them ideal candidates for smart power systems. However these complexities are such that most monolithic solutions are designed around very specific applications usually for high-voltage low-current applications [19].

For more general bridge configurations, the power devices are separate chips and the logic controls are built as a driver unit with suitable output. If appropriately packaged, such a unit operates effectively as a complete bridge control system and the design has small flexibility in the choice of the power control devices (A ... B). Because of the difficulty of driving high-voltage devices, the control part of the system itself has to deal with voltages in the range 50–750 V as this actually contains many of the technologies and designs discussed above to make a high-voltage IC [20, 21].

4.6 CONCLUSIONS AND FUTURE PROGRESS

After a slow start, the development of smart power products has become a reality. Designers and technologists have become familiar with the particular challenge of these new products, and as a result are developing wider ranges of products. Many units are specifically made for large scale users, but gradually modifications of these particular designs for general use are emerging.

The packaging of such products calls for particular solutions and it would be expected that the use of adaptations of existing power packages would decrease and more effective solutions emerge. In the longer term, the gradual production of whole power supply systems controlled from simple logic systems is to be expected.

REFERENCES

[1] Milne A D *'MOS Devices' Design and Manufacture* (Edinburgh: Edinburgh University Press)

[2] Appels J A *et al* 1980 Thin-layer high voltage devices (RESURF devices), *Philips J. Res.* **35** 1–3

[3] Bean K E and Runyan W R 1977 *J. Electrochem. Soc.* **124** 5C

[4] Ohashi H *et al* 1986 *Improved Dielectrically Isolated Device Integration by Silicon-wafer Direct Bonding (SDB) Technique (IEDM)* paper 9.1 p 210

[5] Celler G K, Robinson McD and Lisihner D J 1983 *Appl. Phys. Lett.* **42** 99

[6] Ghandi Sohrab K *Semiconductor Power Devices* (New York: Wiley–Interscience) ch 5

[7] Tihanyi J 1988 *Smart SIPMOS Technology. No 1, Siemens Forsch-u. Entwickl-ber Bd.* **17** (Berlin: Springer)

[8] Allen Phillip E and Holberg Douglas R *CMOS Analog Circuit Design HRW Series in Electrical and Computer Engineering* (New York: Holt Rhinehart & Winston) p 209

[9] *TEMPFET Data Sheet* Siemens BTS 131

[10] Allen Phillip E and Holberg Douglas R *CMOS Analog Circuit Design HRW Series in Electrical and Computer Engineering* (New York: Holt Rhinehart & Winston) p 590

[11] Robb S, Sutor J and Terry L 1989 Smart discretes—a new trend in the power MOSFET industry (*Extended Abstracts*) (*Spring Meeting, Electrochem. Soc., Los Angeles, 1989*)

[12] Denner V, Flohrs P and Michel H 1989 A bipolar smart power technology for high voltage applications (*Extended Abstracts*) (*Spring Meeting, Electrochem. Soc., Los Angeles, 1989*)

[13] Fujihira T *et al* 1991 Self isolation NMOS-DMOS technology for auto-motive low side switches (*IEE ISPSD '91 Meeting, Baltimore, May 1991*) p 242

[14] Zisa M 1988 A smart power innovative actuator in automotive field (*ESSCIRC '88*) p 284

[15] Paxman D, Fisher C, Coe D, Lowis R, Thanigasalam M and Macdonald L 1989 PIPS 500. A Power SMOS Based Process for the

Construction of 50 V Intelligent Power Switches (*Extended Abstracts*) (*Spring Meeting, Electrochem. Soc., Los Angeles, 1989*)
[16] Wrathall R S 1985 The design of a high power solid state automotive switch *Power Electronics Specialist Conference (PESC '85)* (*Toulouse, 1985*) p 229
[17] Weber R 1986 BTS 412. A novel smart SIPMOS switch (*PCI Proceedings, October 1986*) p 271
[18] Gough P, Simpson M and Rumennik V Fast switching lateral insulated gate transistor *IEDM Tech. Dig.* 12–86
[19] Cini C and Diazzi C 1989 250 V integrated full bridge with driving stages and logic inputs in multipower BLC technology (*Extended Abstracts*) vol 89 *Electrochem. Soc.* (*Spring Meeting, Los Angeles, May 1989*) Abstract 306 474
[20] Wegener A 1989 A high voltage interface IC for half bridge circuits (*Extended Abstracts*) (*Spring Meeting, Electrochem. Soc., Los Angeles, May 1989*)
[21] Brüning G and Wegener A 1990 Off-line integrated full bridge driver for induction motors (*Proc. ISPD '90, Shinjuku, Tokyo April 1990*). Abstract 3.2.3

Useful Sources

Symposium on High Voltage and Smart Power Devices ed P Shackle Proc. vol 87–13 Philadelphia May 1987. Electrochemical Society Inc.

Baliga B Jayant *Modern Power Devices* (New York: Wiley)

Review: Murari B *Smart Power* (*ESSCIRC '88*) p 276

5

Optical Actuators

J S McKenzie, K F Hale and B E Jones

ABSTRACT

In this chapter, the area of optical actuators will be reviewed. The desirable features of optical fibre sensors and optically based control systems will be discussed, together with the historical background of the subject. Then the various optical actuation schemes that have been developed and reported in the open literature will be discussed in rather more detail. The actuators that are considered here are mainly those that have been shown to do work on a macroscopic scale, for example, operating a process valve. These optical actuators are optically powered or optically driven and usually employ an additional non-electrical power source, typically a hydraulic or pneumatic one.

5.1 OPTICAL FIBRES

Optical fibres have been commercially available since about 1970, after laboratory development of a multimode fibre was first reported by Kao and Hockham (1966). An optical fibre is a flexible dielectric wave guide which allows light, particularly infrared light, launched at one end to be 'piped' to inaccessible areas over large distances with low attenuation. Optical fibres have been used for both communication and sensor applications, the former being the use which has driven development.

In the field of communications, optical fibres are used to transmit mainly digital information over long distances with very low error rates. Over the last 20 years, vast improvements in multimode and monomode fibres, light emitting diode (LED) sources, laser diodes, photodiode light detectors, fibre terminations and components have led to fibre optic communications networks being laid all over the world, even linking continents. The major reasons for the use of optical fibres over electrical cabling are the very much greater bandwidths available at particular infrared wavelengths, lighter weight and extendibility through multiplexing. Other reasons include immunity from electromagnetic interference, security, safety and low attenuation.

5.2 OPTICAL FIBRE SENSORS

The use of optical fibres in sensors offers all the advantages of the communications use, as well as the possibility of intrinsic safety in hazardous environments, and in some cases, higher accuracy and resolution (Dakin and Culshaw 1989). The first optical fibre sensor (OFS) was developed shortly after the advent of the fibre (Kissinger 1967); this was a multi-fibre displacement sensor. Since then, sensors have been developed to measure linear displacement, rotation, force, pressure, sound, temperature, flow, level and many chemical, biological, electrical and magnetic quantities (Dakin and Culshaw 1989).

An optical fibre sensor consists of a light source (LED, laser or filament source) which is coupled to a length of fibre or fibres. Light is transferred to the sensor where the light is encoded according to the value of the measurand. The encoded light is coupled to the return fibre or fibres. The return fibres can be the same ones that initially delivered the light to the sensor. The encoded light is detected by a suitable detector (photodiode or photomultiplier). The output of the detector is decoded to give the value of the measurand (figure 1). The actual sensor can be any one of a number of optical components (extrinsic sensor) or even the fibre itself (intrinsic

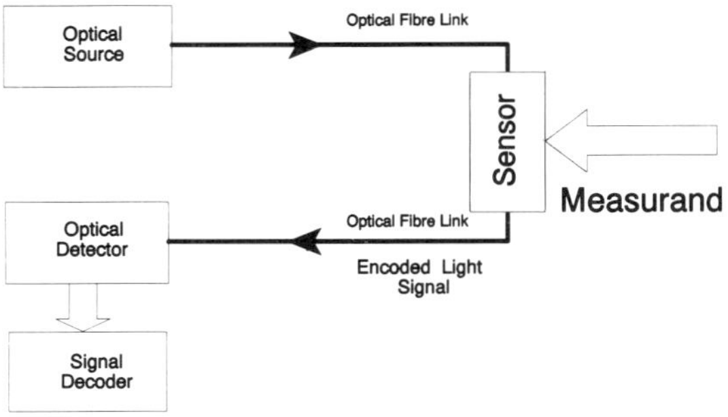

Figure 1. Schematic of an optical fibre sensor.

sensor) providing prospects for distributed sensing. The encoding mechanism can be any one of several optical modulation schemes including intensity, phase, wavelength, polarization and frequency (Dakin and Culshaw 1989). Several measurand signals from different sensors have been sent down a single fibre by employing multiplexing techniques dependent upon wavelength, frequency, time-delay, pulse time or pulse width separation.

Commercial optical fibre sensors are now becoming more widely available (Jones 1985, Jones 1991). As is so often the case, military applications have led the development and optical fibre sensors have been used in weapon systems and more recently combat aircraft. Industrial application has been slow to develop because of device performance limitations, price considerations and to some extent reluctance in accepting new technology.

Sensors often have to be seen in the context of a complete control system where the sensor signal is used as feedback information from the quantity being controlled to the controller.

5.3 CONTROL SYSTEMS

In order to control a quantity effectively, for example displacement of a valve stem or temperature of an oven, it is necessary to know the value of the quantity and take action accordingly. Information about what is happening to the value of the quantity is collected by one or more sensors. This information is used as feedback and when compared to the desired state or set point for a quantity by the controller, an error signal is produced. The error signal is used to drive an actuator, and the actuator works to achieve the set point. Such a system is known as a closed loop control system (Healey 1975). The system under control governs the type of sensors and actuators needed. If the system is one where changes happen slowly, for example certain industrial processes, the sensors and actuators do not need to operate particularly rapidly to keep the process under control. If the system is one where the changes happen quickly, for example, an avionic/flight control system, the sensors and actuators need to respond relatively quickly and be highly reliable to maintain safety of the aircraft.

5.4 ELECTRICAL CONTROL SYSTEMS

Electrical control systems are commonplace in industry today, with a multitude of electrically based sensors and actuators. The controller is typically a complex compensation circuit, more often than not a microprocessor

(Bennett 1988). The actuators can be purely electrically powered units (motors etc) or as in the case of many industrial environments, electrically controlled hydraulic or pneumatic 'safe' power units. A number of companies produce electrical-to-hydraulic, electrical-to-pneumatic or electrical-to-pressure (E/P) converters. These converters take advantage of available hydraulic or pneumatic power supplies to do the majority of the actual work. This power has to be generated, and this can be done in a number of ways depending on the application. It is possible to use hydraulic or pneumatic transmission lines to signal the actuator (powered hydraulically or pneumatically), but over any appreciable distance the time delay introduced would be unacceptable. Electrical transmission lines are significantly faster, and if they are required only to carry the low power needed to operate E/P converters they are safer than the high-power transmission lines necessary to operate large motors. These E/P converters have been commercially available for a number of years and have undergone extensive development (Kranz and Tillmann 1976, Tippetts and Chapin 1976, Taft and Herrick 1981, Zdeblick and Angell 1987). A number of specialist applications use fluidics-based controllers, although these are very rare. The major reasons for the rarity are the speed, cheapness and flexibility of electronics.

5.5 CONTROL-BY-LIGHT SYSTEMS

The benefits of using optical fibre sensors in control systems are not fully realized if it is necessary to provide electrical power or electrical signalling lines for actuation. The controller is, and will be for some time, an electrical/microprocessor-based one. There has been some success in developing an optical computer (Walker 1990), but a suitable production device at reasonable cost is many years away. The interface between an electrical controller and fibre optics is well developed (LED, photodiode, laser etc), and this controller and the interface would be located in a safe environment. Due to the relatively low light attenuation levels of modern optical fibres, typically <3 dB km^{-1} (for light of wavelength 850 nm†) (Cherin 1983), the separation between the interface and the sensors and actuators could be several kilometres. This is likely to be adequate for most applications. Such a system has become known as a 'control-by-light' system and the basic scheme is shown in figure 2 (Collier *et al* 1983). As with electrical actuation systems, it is not practical to have all the actuation power provided down the control signal line, in fact work has been done

† The light attenuation level in an optical fibre is highly dependent on the wavelength of the transmitted light.

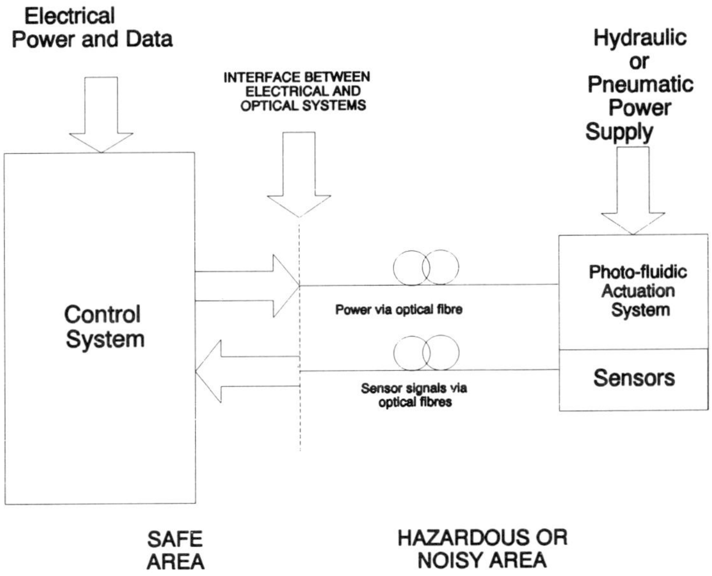

Figure 2. Schematic of control-by-light concept.

which indicates that light levels of the order of 15 mW and greater can be a safety hazard in certain environments (Carleton and Weinburg 1990).

Modern aircraft are increasingly being designed with 'fly-by-wire' controls, which means that a pre-programmed computer system makes all the control adjustments, there being no mechanical link between the controls and the actuators. Many industrial systems are already controlled in this way. Development work is under way to replace 'fly-by-wire' systems by the 'fly-by-light' equivalent using optical fibres instead of electrical cabling (Snitzer 1981, Fox and Snitzer 1982, Baumbick 1985, Glomb 1990). Hence it is desirable to develop and produce optical-to-pneumatic and optical-to-hydraulic converters or photofluidic interfaces to make use of the available 'safe' pneumatic or hydraulic power (Jones *et al* 1983).

5.6 PHOTOFLUIDIC INTERFACES

Light itself is not able to perform any significant amount of mechanical work and needs to be converted to another form of energy in order to do so. This conversion only occurs under certain conditions. When light is incident on an absorbing surface a number of effects can occur depending on the material. Some of these effects can be used to do work either directly or indirectly on a fluid system, the most applicable ones being as follows:

the *photovoltaic* effect: this is the conversion of radiant power into electrical power by means of a p–n junction device;

the *photoacoustic* effect: this is the conversion of modulated radiant power into fluid pressure waves by surface heating of an absorber;

the *photothermal* effect: this is changing the volume an absorber occupies by optical heating due to thermal expansion;

the *photoelectric* effect: this is the generation of free electrons by an incident photon flux from the surface of a metal.

A number of optical-to-pressure converters have been produced using the above effects and these will be examined in the following sections. The actuation schemes have been divided into two sections, based on the manner in which they work employing either photovoltaic based or photothermal/photoacoustic based systems. Optically driven hydraulic or pneumatic systems can therefore be divided into electrically active and electrically passive types.

The active or hybrid system may involve an additional electrical power supply; for example an actuator controlled via an optical fibre data link but powered by a local electrical supply. Such an arrangement could be desirable for a number of reasons, including immunity of the data link from electromagnetic interference. A number of commercial sensors are available with local lithium batteries and optical fibre data links (Jones 1991). For actuation purposes, battery power may not have sufficient longevity, and in general it is necessary to use an electrical supply with the added bulk and an inherent loss of safety. The electrically active sensor or actuator system may be entirely optically powered (Spooncer *et al* 1988), removing the need for an additional power supply.

Only optically powered hybrid systems and non-electrical systems will be reviewed here, as they have more of the advantages of the 'control-by-light' concept.

5.7 PHOTOVOLTAIC-BASED CONVERTERS

5.7.1 Use of Optical Powering

Optical powering of an electrical system was demonstrated at the Bell Telephone Laboratories in 1978 by means of an optically powered piezo-electric tone generator (De Loach *et al* 1978); high optical-to-electrical conversion efficiencies were reported for this system. This work was later extended to the optical powering of circuits to provide two-way speech transmission (Miller and Lawry 1979). In both cases light from a laser diode source was conveyed via a length of optical fibre to a gallium arsenide (GaAs) photovoltaic device. This provided the electrical power required to drive the electrical circuitry.

The optical powering of systems via optical fibres as a means of increasing safety by providing electrical isolation is of interest (Landry *et al* 1991). These power-by-light systems can be used as power sources for actuation or sensor systems (Liu 1991), and conversion factors (electrical-to-optical-to-electrical energies) of 5.5% have been achieved while conversion factors of 12% are expected shortly (Landry *et al* 1991). It is believed that such systems could become available commercially, but with typically lower initial overall conversion efficiencies.

A key element for optically powered devices is the photovoltaic converter. Conversion efficiency with silicon material is between 10 and 15%, but with GaAs material, the figure can increase up to 30%. A prototype 1 mm square GaAs array of eight photoelements produced by Siemens AG (Germany) can provide greater than 60 mW of electrical power via a 10 m long 400 μm optical fibre driven by a laser diode (wavelength approximately 800 nm) and produces about 8 V across a 1 kΩ load (Gross 1991). A 2 mm square multiple array can give a voltage of 32 V. In the near future 10 W high-power laser diode arrays will be available for optical fibre applications. These developments provide prospects for optically powering a wide range of electrical sensors and actuators incorporating a degree of intelligence.

5.7.2 Use of Electromagnetic Devices

In 1981 it was reported that pulse-width modulated LEDs delivering 4–5 mW of electrical power were used to operate an aircraft engine servoactuator well below its designed power rating of 30 mW, resulting in a very slow moving actuator (Snitzer 1981).

An optically powered actuation system was demonstrated at GEC Research (Collier *et al* 1983). Using a silicon photocell, a commercial E/P converter, pneumatic step-up relays and a local pneumatic supply, a control valve was operated (figure 3). Light from a pulsed laser diode was conveyed via 45 m of 200 μm core optical fibre to a solar cell. This had an energy conversion efficiency of about 13%, producing 0–2.2 mA electrical current from the 0–7 mW light power input. The standard E/P converter used normally operates with an electrical input signal of 4–20 mA and produces a 21–103 kPa (3–15 lbf in^{-2}) pressure output. In order to produce a 21–103 kPa signal with just 0–2.2 mA of input current the E/P converter was followed by a pneumatic amplifier stage with a variable offset. The output of this operated a standard 21–103 kPa valve actuator as shown. Later a similar system was operated using a 0–11 mW LED (Collier *et al* 1984).

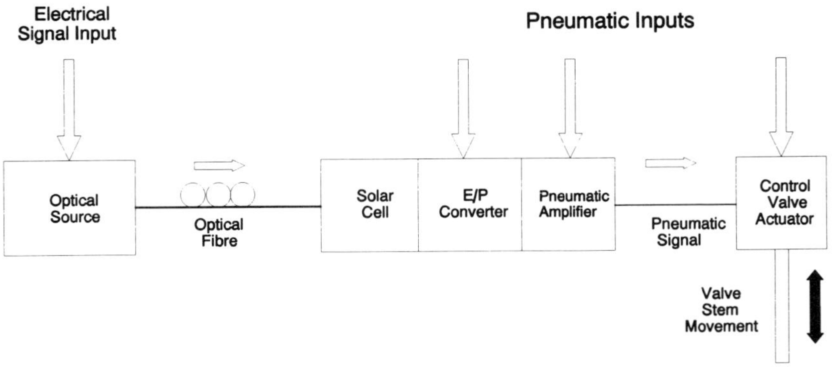

Figure 3. Schematic arrangement of a basic optically-powered system (Collier *et al* 1983).

This method of optical actuation was further developed at UMIST and Brunel University using a commercial E/P converter and a local micropower control system (Liu and Jones 1988, 1989), as shown in figure 4. A standard 4–20 mA electrical signal acts as a set-point control signal; this is converted into a pulse-position modulated light signal, which is transmitted via an optical fibre to a silicon photodiode. The light source used is a CW 30 mW laser diode of wavelength 780 nm. The light power arriving at the

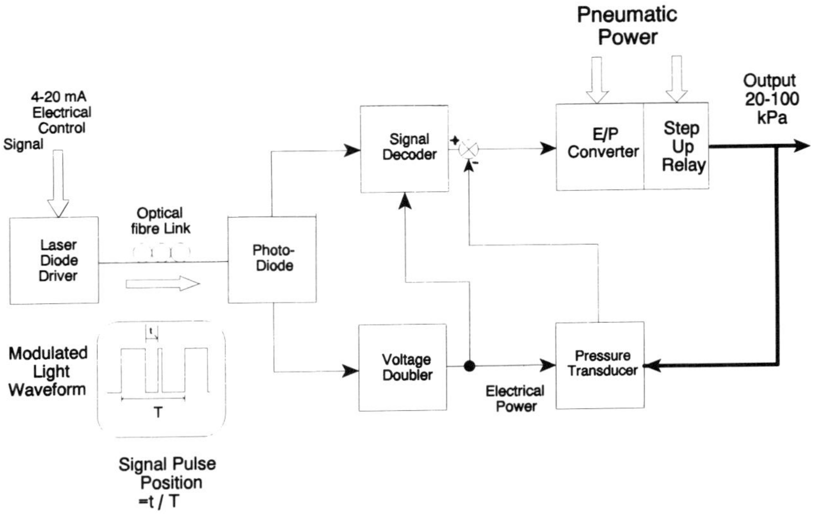

Figure 4. Configuration of an optically-powered precision pressure controller (Liu and Jones 1988).

photodiode averages 8 mW which is converted to 1.2 mW of electrical power by the photodiode. The electrical power produced then goes through voltage doubling and regulation to supply all the electronics in the system. The set-point signal is recovered from the pulse signal by demodulation. A low-power silicon pressure sensor measures the output pneumatic pressure and this information is fed back and compared with the set-point signal. The error signal is then integrated and further amplified to control the E/P converter. The pressure signal from the converter is amplified by pneumatic relays to give a standard 21–103 kPa pressure output. The relationship between the electrical signal (4–20 mA) to the laser-diode driver and the pneumatic pressure (21–103 kPa) is linear to within ±1% and the bandwidth has a value DC to 0.3 Hz. The driver circuit for the laser diode generates a 2.5 kHz square wave and converts the 4–20 mA input current to a 4–20% pulse position which is defined by the ratio of the pulse delay time (t) with respect to the falling edge of the square-wave to the period of the square-wave (T) (inset figure 4).

Later work at Standard Telecommunication Laboratories used much higher launch-light intensity levels of 45 mW, and transmitted the light over two 800 m long 50 μm core optical fibres to power a hydraulic servovalve via two sets of solar cells (Avis and Byron 1987). This corresponds to a transmitted power of 36 mW, allowing for the fibre loss of 1.2 dB km^{-1} at the wavelength of the light source. The output from each optical fibre was incident on four solar cells connected in parallel. Light from a Nd:YAG (wavelength 1060 nm) laser was steered and launched into either one of the fibres by means of a mirror, and depending on which set of cells was illuminated, the servovalve would switch flow from port 1 to port 2 and vice versa. It should be noted that at this wavelength (1060 nm) the response of the silicon photocell is only about 30% of its peak, resulting in a very low optical-to-electrical conversion efficiency. It was reported that semiconductor laser diodes would be used in a practical system.

5.7.3 Use of Piezoelectric Devices

The commercial E/P converters used in the developments referred to above are standard electromagnetic units and the electrical current inputs drive inductive coils in each case. These coils have relatively low resistance and the converters are bulky and require typically about 75 mW of power for full-range operation. These electrical operating powers depend on the device and its intended application. It is possible that a piezoelectric-based converter could require less power, be compact and have a more rapid response, and therefore be more efficient in operation for this application. This alternative approach has been investigated (Liu *et al* 1989).

This system used a photovoltaically driven piezoelectric disc in a differential flapper nozzle arrangement (Liu *et al* 1989), as shown in figure 5. Light from a pulse-width modulated LED (wavelength 820 nm) was launched into 200 m of 200 μm core plastic optical fibre (attenuation 40 dB km^{-1}). A peak optical power of 24 μW irradiates a GaAlAs photocell, giving an output peak potential up to 0.8 V, which is applied to a bi-morph piezoelectric disc. The disc acted as flapper in a differential flapper nozzle arrangement. The outputs of the pneumatic nozzles were fed into the control ports of a Laminar Proportional Amplifier (LPA) gain block.The pressure gain of the LPA block is about 40. The differential output pressures of the LPA block are in the range 35–50 Pa. These amplified output pressures were further stepped up with pneumatic diaphragm amplifiers into the 30–100 kPa range. These pressures are sufficient to drive pneumatic cylinders. This system used significantly lower optical power input than any of the previously described hybrid systems, but there was a significant noise component in the output. The piezoelectric flapper also causes significant zero drift and hysteresis in the system. It is reported that the signal-to-noise ratio and the overall performance of the system would improve with increased optical power input.

The idea of coupling a photovoltaically driven piezoelectric element directly to a LPA to produce a high-performance photofluidic interface was suggested by a group at Cambridge Consultants (Jones *et al* 1990). A schematic of this is shown in figure 6. The 2.2 mW of infrared light, modulated at 1 kHz, emerges from an optical fibre and is converted to 0.045 mW of electrical power by a photodiode (Shelley 1994). The electrical energy passes through a step-up transformer before being fed to a piezoelectric stack. The piezoelectric stack converts the power to a mechanical movement, displacing a volume of oil into one of the control ports of a hydraulically driven LPA and causes a 170 Pa change in pressure. The conversion factor is thus 77 Pa mW^{-1}. After three further LPA stages, each

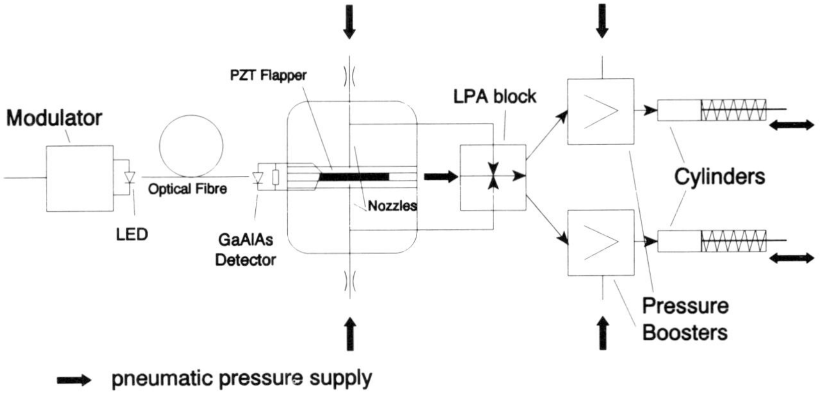

Figure 5. An optically powered piezoelectric flapper optopneumatic converter (Liu *et al* 1989).

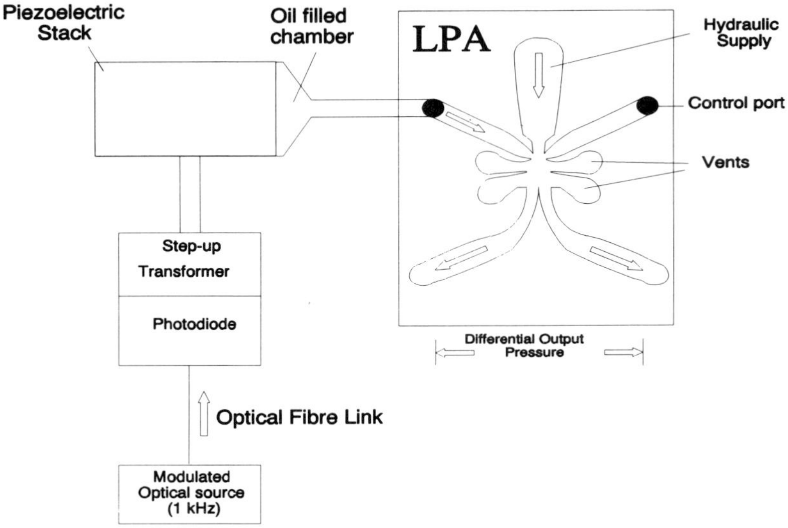

Figure 6. A schematic of a piezoelectrically driven LPA (Shelley 1994).

with a gain of 40, the output signal level is 6.9 kPa (1 lbf in^{-2}). As the system is tuned to a fixed frequency it is necessary in a full working system, to use wave trains of at least a few cycles long. The alternating pressure waves would have to be fluidically rectified and amplified further before being used to operate a conventional servo-type flapper valve. For a working frequency of 1 kHz, the responsiveness of the whole system would be limited to a few hundred hertz which would be acceptable in most application areas.

Another scheme using a piezoelectric element is shown in figure 7 (Hu *et al* 1990). In this arrangement, light intensity modulated at 3.2 kHz is incident on a photodiode. The electrical signal produced causes the piezoelectric disc to oscillate at a similar frequency and this oscillation generates an acoustic signal which is sufficient to switch a turbulence amplifier between the laminar and turbulent flow regimes. The turbulence amplifier switches back when the acoustic signal stops. This switching causes a pressure difference at the output (collector) of the turbulence amplifier. Using 4 mW of optical power, the output pressure had values of 250 Pa and 40 Pa corresponding to the modulated optical signal being off and on respectively. The output pressure signal is of sufficient magnitude to be amplified using commercial pneumatic amplifier relays. This is a digital device, and the switching times are reported to be of the order of 30 ms. It is reported that by using gate-width modulation of the light, a proportional control scheme having a relatively rapid response can be produced.

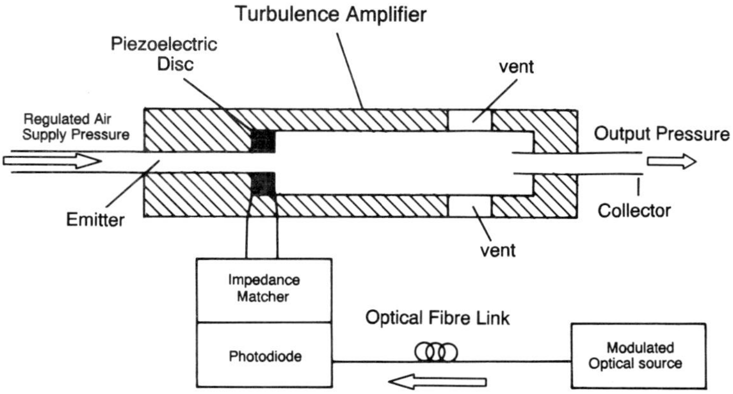

Figure 7. Schematic of a high-speed digital optopneumatic converter (Hu *et al* 1990).

Another version of the piezoelectric switched turbulence amplifier with an acoustic switching frequency of 17.1 kHz had output pressures of 600 Pa (signal off) and 100 Pa (signal on) (Hu *et al* 1991). Again, about 4 mW of optical power was used to power this device. The response time for this pressure change was 10 ms for both rising and falling transients.

The major advantage of this system is that a single low-cost commercial amplifier is needed to drive a pneumatic system without the need for off-setting. This scheme does not require the very clean air supplies necessary for LPA systems, although the air supply must be very carefully regulated to maintain the flow just below the onset of turbulence.

5.7.4 Use of the Photoelectric Effect

An optically controlled silicon microactuator, based on the mechanics of a warped capacitor, has been developed recently (Tabib-Azar and Leane 1990); the basic scheme is shown in figure 8. A potential difference, provided by a battery or power supply, is applied across the plates, one of which is a cantilever beam. The potential difference results in a force which pulls the cantilever beam towards the ground plate. When light of sufficient energy is shone on to the metal ground plate photoelectrons are generated. These migrate through the air gap to the cantilever beam. The photoelectrons reduce the charge on the capacitor, which causes the beam to deflect. A cantilever 600 μm by 50 μm by 1 μm with a gap of 12 μm uses a bias of about 6 V and an optical power flux <1 mW cm^{-2} to actuate 4–5 μm in approximately 30 μs. This miniature actuator requires relatively low levels of light and responds rapidly to light intensity changes, but it is not clear if such a device could be used as a photofluidic interface. The scheme is

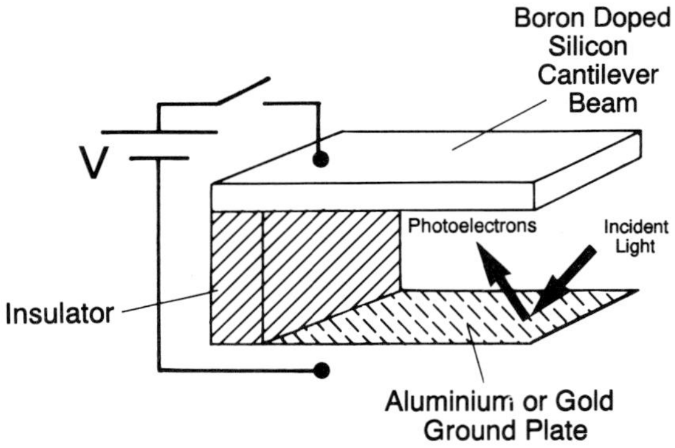

Figure 8. An optically controlled silicon microactuator (Tabib-Azar and Leane 1990).

essentially electrostatic. It may be possible to provide the potential difference needed for the cantilever bias by means of a photodiode array, such that the device might be loosely classified as a photovoltaic based system, although directly activated by incident light using the photoelectric effect.

5.7.5 Performance Comparison

To compare the above photovoltaic based light-to-pressure converters, it is necessary to draw together all the information from the various publications. The conversion factor given is for the optical input through to the first pressure output point and neglects any amplification stages. The results of this comparison are given in table 1.

The GEC pneumatic converter system has a high conversion efficiency and requires little pressure amplification. The pneumatic precision pressure controller has an apparently lower overall conversion efficiency when compared to the GEC system and thus requires slightly more amplification. This is due to part of the electrical power developed at the photodiode being used to power a pressure sensor and control circuitry. In both systems the E/P converter used is relatively slow-acting.

The STL system operates a hydraulic servovalve and very little information is provided. The overall optical power could be reduced with more appropriate selection of source wavelength to the photocells' spectral response.

The pneumatic piezoelectric flapper system has a relatively low conversion efficiency and needs large amounts of amplification. The system does however use substantially lower optical powers than any of the others, thus it can be driven using a low-cost LED.

The Cambridge Consultants piezoelectrically driven LPA has an

Table 1. A comparison of photovoltaic-based converters.

		Optical power used	Optical-to-pressure efficiency (Pa mW^{-1})	Frequency response information
GEC (Collier *et al* 1983)	Pneumatic	7 mW	1570	—
Precision pressure controller (Liu and Jones 1988, 1989)	Pneumatic	8 mW	350[a]	DC-0.3 Hz[a]
STL (Avis and Byron 1987)	Hydraulic	36 mW	—	—
Piezoelectric flapper (Liu *et al* 1989)	Pneumatic	24 μW	57	—
Cambridge Consultants (Shelley 1994)	Hydraulic	2.2 mW	77	1 kHz[c]
Piezoelectric turbulence amplifier (Hu *et al* 1991)	Pneumatic	4 mW	125[b]	16 Hz[d]

[a] Complete control system including feedback pressure sensor.
[b] Light input reduces pressure output.
[c] Working frequency of the first stage of the LPA, maximum probably higher.
[d] Digital converter, switching time 30 ms.

extremely high speed of response and a reasonably high conversion efficiency, but as it is based on a laminar device it will suffer from the problem of maintaining laminar flow conditions with changing working fluid temperatures. As a result its performance with changing temperature will be poor for fluids such as standard hydraulic oils. This will be discussed in more detail in section 8.1.

The piezoelectrically switched pneumatic turbulence amplifier has a quite low conversion efficiency, but it is a relatively fast device.

The photovoltaic-based systems described above all provide electrical isolation and immunity of the transmission line from electromagnetic interference, through the use of light and optical fibres. However, as an electrical intermediary is used, the electrical components may still require careful insulation in hazardous environments, and where there is the possibility of electromagnetic interference. In the case of the optical actuators it can be argued that it is more convenient and desirable to convert light power into fluid pressure pulses, using the photoacoustic or photothermal effects.

5.8 PHOTOACOUSTIC- AND PHOTOTHERMAL-BASED CONVERTERS

In 1881, A G Bell used the photoacoustic effect to produce sound in a gas with a chopped beam of sunlight, and this was later used to produce the photophone (Mimms 1982). It sent 'voices' through the air, but it was not a practical system because rain, snow and other obstructions can block light. For most of the century following this discovery, the photoacoustic effect was used only to study the absorption of light by materials for analysis known as photoacoustic spectroscopy (Rosencwaig 1980). In 1976 the fibre optic photophone, a two-way optical telephone system employing a photoacoustic receiver, was demonstrated at Bell Telephone Laboratories (Nelson *et al* 1976). The use of fibres eliminated the problem of obstructions encountered by Bell.

5.8.1 Use of Laminar Proportional Amplifiers

A combined photoacoustic and fluidic actuator system was demonstrated at the Harry Diamond Laboratories (HDL) (Harry Diamond Laboratories 1980). Further details of the system, developed primarily for aerospace applications, were released in 1984 (Gurney 1984). At the heart of the system was a modified LPA. One of the control ports of this modified LPA contained an open photoacoustic cell of volume ~7 mm^3, and the other control port was grounded (figure 9). Light, square-wave modulated at 1400 Hz, passed through a plexiglass cover window into the cell and was incident on a carbon black absorber. An acoustic signal of the same frequency resulted. The acoustic pressure pulses were amplified by the interface stage and two subsequent LPA stages. The total gain of these three stages is 480. The next section of the photofluidic interface is a fluidic rectifier which doubles the input frequency and delivers an average output pressure which is proportional to the input pressure amplitude. The DC output of the rectifier is further amplified by two more LPA stages and diaphragm amplifiers, to a level sufficient to actuate a piston. The LPAs used were standard C-type laminates, pneumatically driven. The optical source

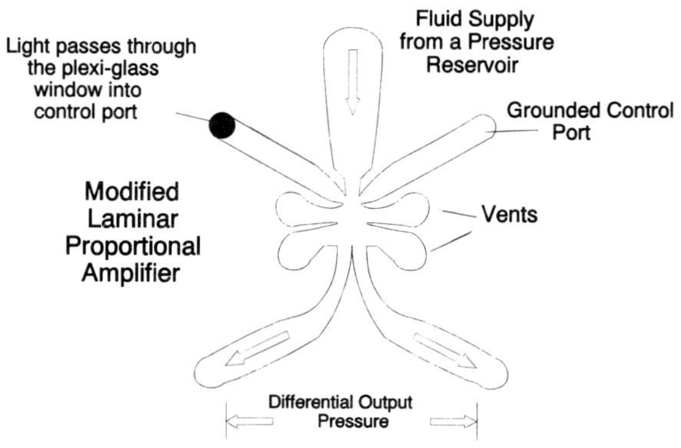

Figure 9. A LPA photofluidic interface (Gurney 1984).

was a helium–neon gas laser which produced up to 9 mW of optical power (peak-to-peak) at the cell. The incident light produced a pressure approximately 0.04 Pa mW^{-1} at the cell and a pressure of approximately 0.4 Pa mW^{-1} after the interface. It was reported that the output of the rectifier could be controlled by modulating the light input in the following ways: pulse modulation, frequency modulation, gate-width modulation, amplitude modulation, and the last two were the most successful schemes. The complete photo-fluidic system threw a bell crank full travel at 7 Hz or less. The system was also operated using a 2.5 mW, 850 nm laser diode.

The work was extended to produce a hydraulic version of the interface (Bartholomew 1984, Drzewiecki *et al* 1985). Further improvements were made by using higher input impedance LPAs, and this had the expected effect of increasing the pressure developed at each stage (Gillespie 1990). Using MIL-H-5606 hydraulic oil and 10 mW of light, the improved interface and revised amplifying stages were able to generate a pressure of 138 kPa, sufficient to drive an actuator or servovalve. The time constant for this system was 8 ms. A large number of LPAs were required to amplify the small signal generated making this an expensive device.

This type of photofluidic interface also forms part of a proposed fluidic-based kinetic energy weapon for intercepting hostile missiles from space (Drzewiecki *et al* 1991). A laser beam mounted on the launch vehicle relays commands to the projectile (to adjust its course) these are received by the photofluidic detector on the cell. The command signals are amplified and processed using gas driven fluidic components to activate one of four thrusters; the device also incorporates a fluidic inertial guidance system. The photofluidic interface uses a much higher light modulation rate (4.4 kHz) indicating that the interface is faster than the original (Gurney 1984). The laser activated thruster unit with fluidic sensors has been constructed and tested on the ground (Drzewiecki *et al* 1991).

The use of this type of photofluidic interface coupled with optical fibre sensors for vibration control in smart structures has been proposed by Claus *et al* (1991). In this scheme, optical fibre sensors attached to (or embedded in) the structure replicate the structural vibrational shapes and modes, and the output signal is fed into an active optical amplifier which raises the level of the optical output signal. This is then coupled into a photoacoustic cell and amplified by an LPA and, after signal processing (amplification etc), the output pressures will act on the vibrating structure to control the vibration. The complete control system has not as yet been constructed (Claus *et al* 1991).

Another LPA-based photofluidic or optofluidic interface was developed at the United Technologies Research Centre (UTL) (Hockaday *et al* 1984). The interface was quite different from that developed at HDL in that this interface used a modified LPA without control ports. As with HDL interface, the top cover was replaced with an optically transparent window.

The rear wall of the LPA was composed of a graphite–epoxy composite which acted as a light energy absorber (figure 10). The light from an argon ion 10–30 mW output laser was focused by a microscope objective lens to a 15 μm spot on the absorber, located in an expanding section of the power (supply) jet. The graphite–epoxy absorber transferred the energy from the laser via heat to the sensitive boundary flow layer, situated in the power jet of the modified LPA. This had the effect of altering the balance of flow between the two output ports, which caused an optically induced pressure change. A fluidic gain block was built with six LPA stages. The gain block had two sections, a preamplifier with an amplification of 13.75 times and a power amplifier with a further gain of 20 times. When driving a blocked load, conversion factors for this pneumatic system were dependent on the interface supply pressure; these were quoted as 15.5 Pa mW^{-1} for a 20.7 kPa supply and 10.9 Pa mW^{-1} for a 13.8 kPa supply, both for up to 25 mW optical power input. The bandwidth was quoted as DC to 140 Hz. A closed-loop pneumatic control system was demonstrated using this interface (Hockaday and Waters 1987).

Following on from the pneumatics demonstrator at UTL, a hydraulic version of the interface was developed. The hydraulic interface was operated using a noiseless hydraulic source to minimize errors caused by hydraulic pump noise. Using MIL-H-5606 oil, the conversion factor (again dependent on the interface supply pressure) was quoted as 44 Pa mW^{-1} for a 690 kPa supply pressure. This figure was for up to 25 mW of optical power, again when the fluidic system was driving a blocked load. The bandwidth for the hydraulic system was reported as DC to 145 Hz (Hockaday and Taylor 1986). Further developments of this hydraulic interface resulted in an improvement in the conversion factor to a value of 113 Pa mW^{-1} for a

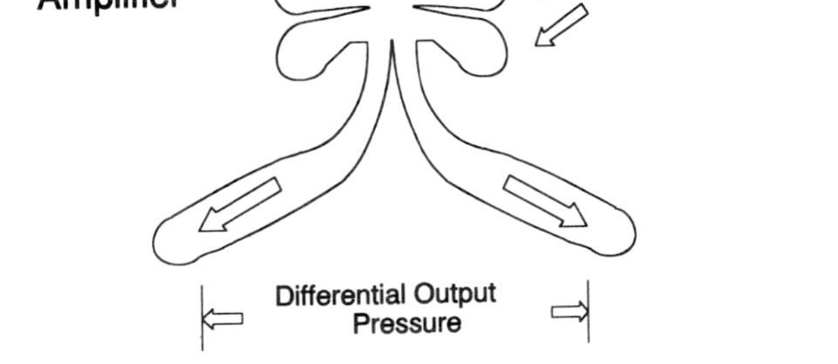

Figure 10. A LPA optofluidic interface (Hockaday *et al* 1984).

896 kPa supply pressure and the bandwidth to 170 Hz (Hockaday 1989). Recent results have shown that by increasing the supply pressure still further to 1300 kPa the conversion factor increases to 170 Pa mW^{-1} (Hockaday and Waters 1990). This interface is being considered by the American military for use in the next generation of aircraft (Glomb 1990).

A problem with the UTL interface is the sensitivity to the position of the focused light spot. It was reported that moving the 15 μm spot from the optimum position by 10 μm, normal to the edge of the nozzle caused the pressure response to drop by a factor of two (Hockaday *et al* 1984). The response is slightly less sensitive to displacement along the wall. The light spot position is clearly critical to the interface performance, thus the interface does require some very precise, stable and probably expensive optical components. The exact mechanism of the optically induced deflection of the supply jet was not revealed until recently; it is due to the optical modulation of the fluid viscosity. This creates an asymmetric jet flow which propagates towards the splitter (Hockaday and Waters 1990). An analytical model for this system based on the optical production of a viscosity gradient has been developed (Dohta *et al* 1990, Dohta *et al* 1991). This model predicts the relationship between the differential output pressure, optical input power, the optical spot size and position (normal to the power jet).

The operating principle of the Hockaday photofluidic interface was applied to the switching of an attached jet by laser light (Dohta *et al* 1991). This basis of the interface was a wall attachment fluidic device similar to the LPA but with slightly different geometry; the device output is essentially digital and it has a memory for its last state. By focusing the laser beam to a spot on one side of the power jet, it was possible with a minimum of 15–20 mW to switch the output to the opposite wall. The output could be switched back by applying a minimum of 15–20 mW to a different spot on the other side of the power jet. The position of the spots within the power jets affected the amount of optical power required to switch the jet. The pneumatic supply pressure used was 340 Pa and the pressure developed at one of the outputs is expected to be slightly less than this.

All the LPA-based converters suffer from the problem of maintaining laminar flow conditions with changing working fluid temperatures. This requires the operating pressure to vary (depending on the working fluid) by a significant amount. It is possible to compensate, to a limited extent, for the large changes in supply pressure using temperature-compensated regulators and temperature-sensitive restrictors (Cycon and Shaffer 1982). This does however increase the complexity and cost of the systems considerably.

5.8.2 Use of a Closed Cell

In 1988 a group working at Brunel University developed an optopneumatic converter using a closed optoacoustic cell and a flexible membrane-nozzle arrangement (Hale *et al* 1988). A small amount of air is enclosed in a miniature cylindrical cell (approximately 5 mm^3 in volume) which has a thin elastic membrane stretched across one end (figure 11). An optical fibre is introduced into a hole in the opposite face. Inside the cell is a carbon-black-on-filter-paper absorber. The energy from the light causes the absorber, and consequently the air in the cell to heat and expand. This expansion in turn causes the flexible membrane to move outward. The membrane movement is detected by a change in back-pressure of an impinging jet of air that emanates from a nozzle. A differential configuration is used in order to produce a device which responds only to differential pressure changes in the two cells. The cells are situated close to one another so that ambient changes do not introduce errors. This change in back-pressure is amplified by a five-stage fluidic gain block. The resulting differential pressure is put through a fluidic flip-flop device. The output of the flip-flop is further amplified using commercial pneumatic step-up relays. This operates a spool valve and actuates a double-acting pneumatic cylinder. The light is produced by a pulsed 850 nm wavelength laser diode and the optical power averages 5 mW at the cell. Before amplification a direct differential pressure change of 150 Pa is produced in 0.5 s, 250 Pa in 1 s and 300 Pa in 2 s. Up to a response time of 0.5 s the conversion factor has a value of 60 Pa mJ^{-1}. For the 300 Pa produced after 2 s the light power/pressure conversion factor is 60 Pa mW^{-1}. The conversion factors for this device are time dependent, implying the converter is an integrating device. The whole system operates on/off in approximately 300 ms. The system has operated successfully with up to 30 m of 600 μm core plastic-coated glass fibres.

An incremental control system using the optopneumatic converters developed at Brunel University has been demonstrated (Hale *et al* 1990, 1991); it is capable of operating a standard 21–103 kPa (3–15 lbf in^{-2}) valve actuator (figure 12). A differential configuration of the optopneumatic converter cells is also used in this control scheme. The differential pressure

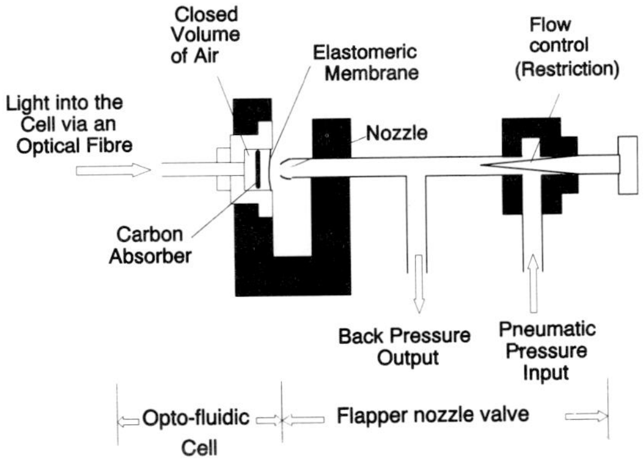

Figure 11. A closed-cell optopneumatic converter (Hale *et al* 1988).

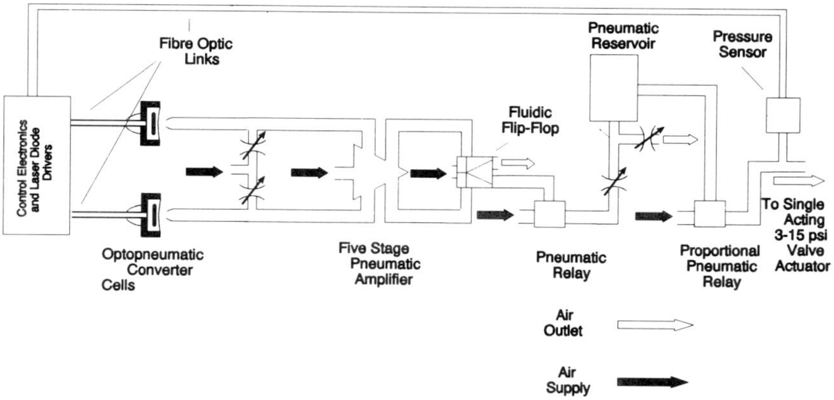

Figure 12. Optopneumatic incremental control system (Hale *et al* 1990/1991).

change can be amplified if need be by a fluidic gain block so that it can operate a pneumatic flip-flop. One of the outputs of the flip-flop is vented to the atmosphere and the other output is connected to a pneumatic step-up relay. The output from the step-up relay is able to pump up a leaky reservoir. The pressure in the reservoir powers a proportional relay which provides the air pressure/air flow amplification between the reservoir and the standard actuator and valve. The pressure is monitored by a pressure sensor, the output of which is compared with a selected set-point. A simple comparator is connected to the laser diode driver circuits so that either of the laser diodes can be operated, depending upon whether the final output pressure is required to increase or decrease.

Using a CW 780 nm wavelength laser diode which produces an average of 3.5 mW optical power at the cell, a direct differential pressure of 500 Pa is produced in 0.5 s, and a maximum of 1400 Pa in 5 s. Within the first 0.5 s, there is a conversion factor of 286 Pa mJ^{-1}. After 2 s, the light power/pressure conversion factor is 160 Pa mW^{-1}, and after 5 s the conversion factor is 400 Pa mW^{-1}. With the improved sensitivity it is possible to dispense with the fluidic gain block, although when it is included the speed of response for the system does improve. The time response for each 6.9 kPa incremental change is of the order of 1.5 s, with the fluidic gain block, and 2 s without it.

The speed at which the Brunel device operates is adequate for some industrial applications; many applications, however, need a quicker speed of response. This has been achieved by reducing the thermal inertia of the system and, for pneumatic systems, by reducing the fluid volumes in the nozzle (McKenzie and Clark 1992). A revised schematic is shown in figure 13. As before, light from a laser diode was conveyed via an optical fibre to the closed cell. The cell contains a fibrous absorber and air. The square cells

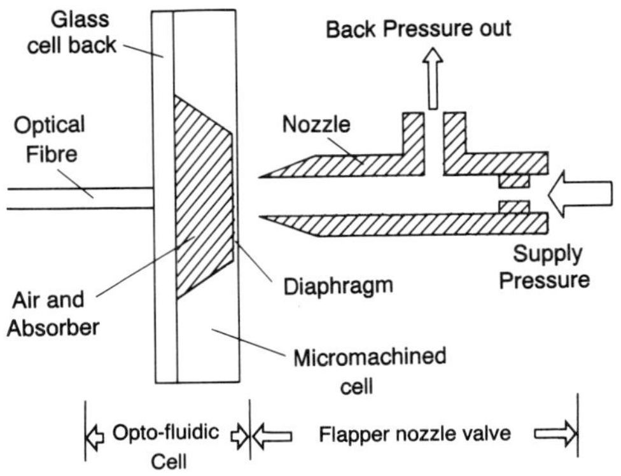

Figure 13. Schematic of the improved Brunel converter (McKenzie and Clark 1992).

with integral diaphragms were micromachined from a silicon wafer; the cell volumes were about 0.9 mm^3. The circular-section nozzles were produced using standard machining techniques.

The improved converter can be up to 35 times faster than the original, with a pressure rise time of 21 ms when pneumatically driven. The best rise time for the hydraulic converter was 67 ms (McKenzie and Clark 1992). The best sensitivity achieved by the improved converter was 304 Pa mW^{-1} when pneumatically driven and 75 Pa mW^{-1} when hydraulically driven.

The speed of operation of the converter has been improved substantially as a result of reducing the physical dimensions of the device and the thermal inertia of the cell contents. The device may be operated using a standard hydraulic oil (the rubber membranes used previously were attacked by oils). Micromachining also allows much more closely matched pairs to be produced than was possible with the rubber membranes device; there is a corresponding greatly reduced sensitivity to ambient temperature change. A further development of the device has been the use of silicon micromachining techniques to produce adjacent pairs of cells and pairs of square micromachined nozzles with the nozzle/membrane separations defined as a recess in the cell pairs (McKenzie 1993). A schematic of the assembled device (produced in adjacent pairs) is shown in figure 14.

Another device employing miniature cells is a micropump intended for medical applications (Mizoguchi *et al* 1992). This device employs an array of closed cells which will be optically driven. The cells with integral pre-deflected membranes were micromachined in silicon (volume = 0.25 mm^3) and filled with Freon 113. A carbon wool absorber converted the light into heat in the cell. Using these cells, large deflections have been obtained when the Freon 113 undergoes a phase change (liquid to vapour). It has been found that each cell requires about 11 mW of power to operate. The array of

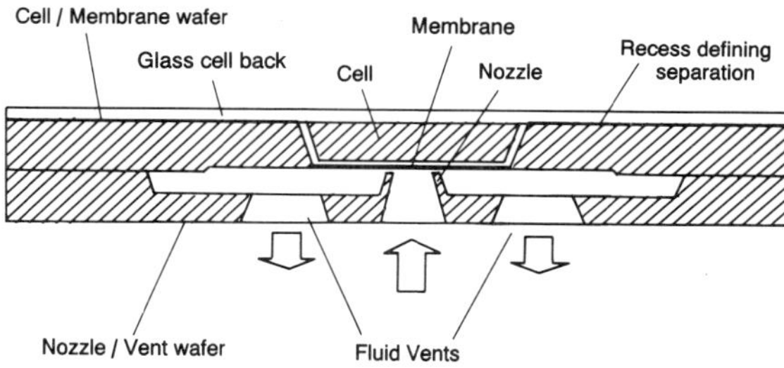

Figure 14. Schematic of the cell and nozzle wafers bonded together (McKenzie 1993).

cells was covered by a thin glass plate with an 18 μm gap which made a flow channel (see figure 15). The cells were operated in sequence in a peristaltic manner. The micropump cells have, so far, only been driven using small air pressures (no Freon and absorber in the cells), and in this manner the micropump has pumped small quantities of water. However, it is expected that when the cells are optically driven the micropump will be capable of pumping water and the operating frequency of the pump is expected to be approximately 3 Hz. Typical fluid pumping displacements are approximately 30 nl per cycle. The operating temperature range is obviously limited due to the phase transition temperature, but this is not a problem for the intended application in the medical area.

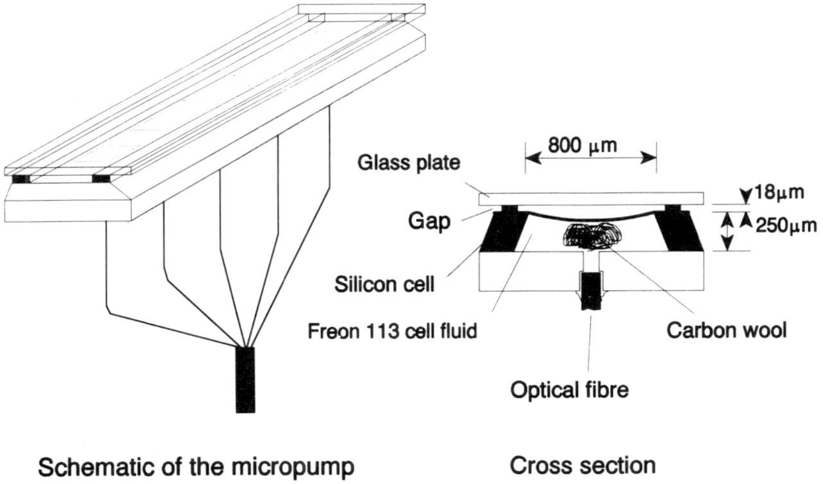

Figure 15. Schematic of the optical micropump (Mizoguchi *et al* 1992).

5.8.3 Other photothermal devices

The Furukawa Electric Company developed an optically activated switch (Jones *et al* 1983). Light from a high-powered laser, via an optical fibre, heated a bimetallic element; when the element was heated through 60 °C the switch opened, and when the light was switched off the switch closed.

A direct optical actuation scheme was reported by researchers at STC Technology (STL) (Hale and Jones 1986). A modified thermostatic valve was operated optically in five seconds by expanding a wax-in-copper capsule (figure 16). The total movement was reported to be 1 mm; there was no indication of the light power used to achieve this actuation. The system would have had a very narrow operating temperature range due to the design criterion for thermostatic valves. Apparently, no further work beyond the simple prototype was conducted.

A light sensitive polymer gel has been reported by workers at MIT (Watts 1990). When light is shone onto the gel it shrinks to as little as one thousandth of its size with 100 mW of 488 nm wavelength laser light at room temperature. The heating effect of the laser light causes the polymer to change state or phase; one phase of the polymer takes up less space than the other, hence it shrinks. For a 100 μm ball of the gel a change between the two phases takes about 1 s. It is unlikely that this gel could be used as a photofluidic interface unless the phase change temperature is above the working conditions of the interface. It is possible that this is the case.

A number of researchers have worked on optically excited structures for sensor applications; these tend to be resonant devices producing frequency output (Hok *et al* 1989). The sensors are usually micromachined devices which are optically excited and interrogated and have been investigated to measure pressure (Uttamchandani *et al* 1987), force (Mallalieu *et al* 1985) and temperature (Thornton *et al* 1986). The resonant frequency of the given structure changes with the measurand. Most of these devices involve small

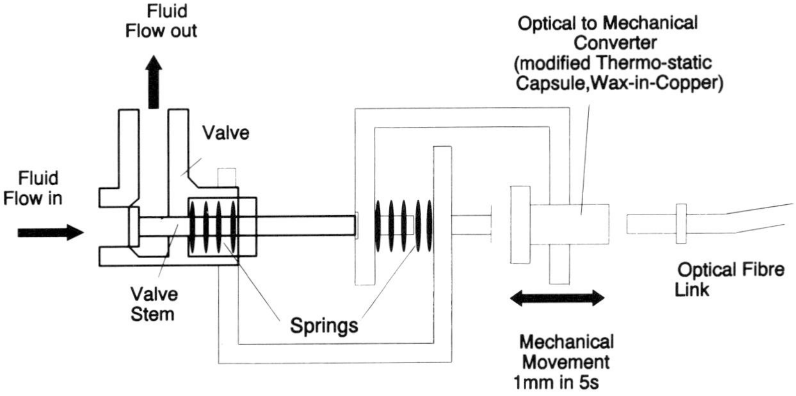

Figure 16. Photothermal conversion scheme (Hale and Jones 1986).

(sub-micron) movements and it is unlikely that any of these devices could be used as photofluidic converters. In 1986 a resonance pressure sensor was reported using an all fibre optical actuator (figure 17). It employed a thin glass tube pressure chamber which shifted its resonant frequency with changing chamber pressure. The movement of the tip was quoted as 20 μm, sufficient to be detected by a moving shutter between the two fibres (Johnson 1986).

5.8.4 Performance Comparison

In order to compare the above photoacoustic/photothermal optical actuation schemes it is necessary to draw together all the information elicited from the publications under as uniform a set of conditions as possible. This presents quite a problem considering that all the converters work in such varied ways. All the systems will be considered on the basis of the interface device. For the LPA-based devices this is considered to be after the first stage of a modified LPA and for the closed-cell optopneumatic converter, the differential back-pressure output at the flapper nozzles; in all cases the converters are driving a blocked load (e.g. a pressure transducer etc). The results of this comparison are given in table 2.

The HDL pneumatic converter has a relatively low conversion efficiency, thus to step the pressure up to usable levels requires a large number of LPA stages. With each LPA stage, there is an associated time delay in signal propagation. This, when coupled with the need to average the AC signal to produce a DC level, accounts for the difference between the first-stage

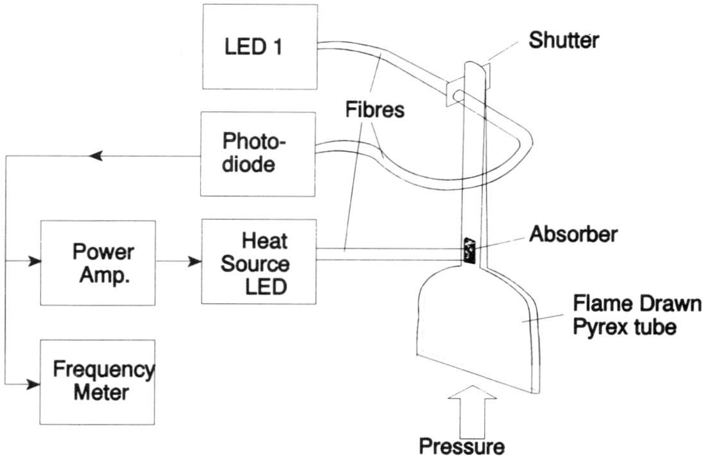

Figure 17. Resonance pressure sensor using an optical actuator (Johnson 1986).

Table 2. A comparison of photoacoustic/photothermal based converters.

		Conversion factor (Pa mW^{-1})[a]	Frequency response of the first stage (Hz)[a]	Bandwidth of whole systems produced (Hz)
HDL	Pneumatic (Gurney (1984)	0.4	30–1400[b]	7[c]
	Hydraulic (Gillespie 1990)	>0.4	30–1400	>39[c]
UTL	Pneumatic (Hockaday and Waters 1987)	15.5	DC-140	3[d]
	Hydraulic (Hockaday and Waters 1990)	170[d]	DC-170	Unknown
Brunel	Pneumatic (Hale *et al* 1990)	400	DC-1/10	≃1
	Hydraulic	—	—	—
Brunel (miniature)	Pneumatic (McKenzie and Clark 1992)	304	DC-40	—
	Hydraulic (McKenzie and Clark 1992)	75	DC-15	—

[a] All driving a blocked load.
[b] Resonance peak.
[c] Signal processing done in the frequency domain due to zero DC response.
[d] Data at 21 °C.

frequency response and the whole system bandwidth. The hydraulic version is more efficient due to better impedance matching of the photoacoustic signal and the LPA stage.

The UTL pneumatic converter also has a reasonably low conversion efficiency and thus requires a number of LPA stages. It is expected that the frequency response of systems produced will be substantially greater than the figure quoted when more appropriate flow amplification is used. The hydraulic version is more efficient than the pneumatic converter, due to more effective heat transfer from the absorber to the hydraulic oil and higher supply pressures. It is therefore expected that fewer amplifier stages will be required and thus the frequency response of any systems produced will be greater than the pneumatic system.

The Brunel pneumatic converter (with rubber membranes) has a high conversion factor and thus requires less amplification. However, this is an integrating device and this high conversion efficiency is at the expense of the speed of operation. This converter was aimed at an industrial application where this response is often adequate. The miniature version of this device which is capable of being operated pneumatically and hydraulically offers a greater speed of response which may open up new application areas such as aerospace.

5.9 CONCLUSION

It is argued that the logic of installing optical fibre sensors on account of their freedom from electromagnetic interference and of their safety in hazardous atmospheres is discounted if electrical power is still needed to relay commands to actuators. This belief has stimulated the development of optically powered fluidic actuators, and an investigation of non-electrical optofluidic interfaces using the photoacoustic/photothermal effects in closed cell or laminar proportional amplifiers. These developments may result in the production of 'control-by-light' pneumatic and hydraulic systems.

ACKNOWLEDGMENTS

The authors wish to thank the Science and Engineering Research Council (now EPSRC), OSCA and the various industrial companies for support of the research work at UMIST and Brunel University.

REFERENCES

Avis S E and Byron K C 1987 Remote actuation by optical power transmission in fibres *Optics and Laser Tech.* **19** 102–104

Bartholomew R D 1984 Optical fluidic servo-valve *Report No SUAAVRADCOM TR 83 35* (Boeing Military Aircraft Company)

Baumbick R J 1985 Fibre optics for propulsion control systems, *Trans. ASME: J. Eng. Gas Turb. Power* **107** 851–855

Bennett S 1988 *Real Time Computer Control* (New York: Prentice Hall)

Carleton F B and Weinburg F J 1990 Radiative ignition by loose agglomerates of fine fibres *OSCA Doc. No.90/56* (Chislehurst, Kent: Sira Ltd) 1–10

Cherin A H 1983 *An Introduction to Optical Fibres* (Singapore: McGraw Hill)

Claus R O, Murphy K A, Fogg B R, Sunn D and Vengsarkar A V 1991 Photofluidics for integrating fibre sensors with high authority mechanical actuators, *SPIE (Fibre Optic Smart Structures and Skins)* **1588** 159–168

Collier M J, McGlade S M and Stephens P E 1983 The optical actuation of a process control valve, *IEE Conf. Pub. No 221 1st Int. Conf. Optical Fibre Sensors* 57–61

Collier M J, McGlade S M and Stephens P E 1984 The optical actuation of a process control valve, *GEC J. Res.* **2** 125–128

Cycon M F and Shaffer D J 1982 Design guide of laminar flow fluidic amplifiers and sensors, *US Gov Report No HDL-CR-82-288-1*

Dakin J and Culshaw B (eds) 1989 *Optical Fibre Sensors: Systems and Applications* vols 1 & 2 (Boston and London: Artech House Inc)

De Loach B C, Miller R C and Kaufman S 1978 Sound alerter powered over an optical fibre, *Bell Syst. Tech. J.* **57** 3309–3316

Dohta S, Matsushita H and Takamori T 1991 A study of photofluidic interface (*Fluecom '91 3rd Triennial Int. Symp. on Fluid Control, Measurement and Visualisation, New York, ASME, 1991*) 193–197

Dohta S, Tokai M and Takamori T 1990 Studies on photofluidic interface, *Trans. Soc. Instrument and Control Engineers* **26** 780–786

Drzewiecki T M, Holmes A B, Packard E A and Kiernan R J 1991 Fluidic guidance for space based kinetic energy weapon projectiles (*Fluecom '91 3rd Triennial Int. Symp. on Fluid Control, Measurement and Visualisation, New York, ASME, 1991*) 303–311

Drzewiecki T M, Toda K and Gurney J O 1985 Photofluidic interface US Patent 4512371

Fox E V and Snitzer E 1982 Fly-by-light sensors, *Proc. SPIE Fibre Optics* **355** 148–157

Gillespie G C. 1990 Private communication (Maryland, USA: Harry Diamond Laboratories)

Glomb W L 1990 Fly-by-light schemes move into the demonstration stage, *Laser Focus World* **June** 1990 167–176

Gross W 1991 Optical power supply for fibre-optic hybrid sensors, *Sensors and Actuators* **A25–A27** 475–480

Gurney J O 1984 Photofluidic interface, *Trans ASME J. Dynamic Systems, Measurement and Control* **106** 91–97

Hale K F, Clark C, Duggan R F and Jones B E 1988 High sensitivity optopneumatic converter, *Proc. IEE Part D* **135** 348–352

Hale K F, Clark C, Duggan R F and Jones B E 1990/1991 Incremental control of a valve actuator employing optopneumatic conversion
1990 *Sensors and Actuators* **A21–A23** 209–210
1990 *Proc. Int. Sym. on Fluid Power* Cambridge UK 103–109
1991 *Inst. J. Optoelectronics* **6** 539–544

Hale P G and Jones R E 1986 Optical fibre systems for offshore monitoring and control, *SPIE Fibre Optics* **630** 239–243

Harry Diamond Laboratories 1980 Optofluidic interface transducer, *Technical Sheet FC-107* (Maryland, USA: Harry Diamond Laboratories)

Healey M 1975 *Principles of Automatic Control* (London: Hodder and Stoughton)

Hockaday B D 1989 Optical actuation for aircraft control, *OSCA Doc. No 89/95* F (Chislehurst: Sira Ltd) 1–11

Hockaday B D, Glomb W, Taylor K and Waters J 1984 A novel opto-fluidic interface, *Proc. Am. Control Conf.* (*San Diego, June 1984*) 1038–1040

Hockaday B D and Taylor K D 1986 An optical to hydraulic interface, *Proc. Am. Control Conf.* (*Seattle, June 1986*) 2048–2051

Hockaday B D and Waters J P 1987 Opto-mechanical smart actuator, *Proc. Am. Control Conf.* (*Minneapolis, June 1987*) 735–736

Hockaday B D and Waters J P 1990 Direct optical-to-mechanical actuation, *Appl. Optics* **29** 4629–4632

Hok B, Tenerz L and Gustafson K 1989 Fibre optic sensors: a micromechanical approach, *Sensors and Actuators* **17** 157–166

Hu F Q, Watson J M, Page M and Payne P A 1990 A high speed opto-pneumatic digital actuator, *Proc. Conf. Appl. Optics and Optoelectronics* (*Nottingham, September 1990*) ed J C Dainty (Bristol: Institute of Physics) 97–98

Hu F Q, Watson J M, Page M and Payne P A 1991 A fast optopneumatic converter for robot actuation, *IEE Colloquium on Robot Actuators* Digest No 1991/46

Johnson M 1986 Optical actuator frequency coded pressure sensor, *Optics Lett.* **11** 587–589

Jones B E 1985 Optical fibre sensors and systems for industry, *J. Phys. E: Sci. Instrum.* **18** 770–782

Jones B E 1991 Optical fiber sensors for industrial measurement, *Revue Française de Mecanique* No 1991-2 ISSN 0373-6601 133–138

Jones B E, Beck M S and Watson J M 1983 Feasibility study of opto-pneumatic converters, *OSCA Doc. No 83/88F* (Chislehurst, Kent: Sira Ltd) 1–49

Jones B E, Kerry N J and Hazel M S 1990 An analysis of photomechanical transduction processes. *Proc. Conf. Applied Optics and Optoelectronics*

(*Nottingham, September 1990*) ed J C Dainty (Bristol: Institute of Physics) 95–96

Kao K C and Hockham G A 1966 Dielectric fibre surface waveguides for optical frequencies, *Proc. IEE* **117** 1151–1158

Kissinger C D 1967 Fibre optic proximity probe US Patent 3327584

Kranz W and Tillmann H 1976 Electrofluidic converter US Patent 3942559

Landry M J, Rupert J W and Mittas A 1991 Power-by-light systems and their components: an evaluation, *Appl. Optics* **30** 1052–1061

Liu K 1991 Power budget considerations for optically activated conventional sensors and actuators, *IEEE Trans. on Instruments and Measurement* **40** 25–27

Liu K and Jones B E 1988 Optoelectronic control of pneumatic pressure using a local micropower feedback sensor, *Proc. Conf. Sensor '88 (Nuremberg, May 1988)* part B 297–313

Liu K and Jones B E 1989 Pressure sensors and actuators incorporating optical fibre links, *Sensors and Actuators* **17** 501–507

Liu K, Jones B E and Watson J M 1989 Optoelectronic actuation using a piezoelectric flapper-nozzle arrangement, *Proc. Conf. Sensors and their Applications IV* (*Canterbury, September 1989*) (Bristol: Institute of Physics) 68–69

Mallalieu K, Youngquist R and Davies D E N 1985 An analysis of the photothermal drive of a quartz force sensor, *SPIE* **522** 211–219

McKenzie J S 1993 Optical actuation and control-by-light systems PhD thesis, Brunel University UK

McKenzie J S and Clark C 1992 High sensitivity micromachined optical-to-fluid pressure converter for use in an optical actuation scheme, *J. Micromech. Microeng.* **2** 245–249

Miller R C and Lawry R B 1979 Optically powered speech communication over a fibre light guide, *Bell Syst. Tech. J* **58** 1735–1741

Mimms F M 1982 The first century of light wave communications, *Int. Fibre Opt. Comm.* **3** 10–26

Mizoguchi H, Ando M, Mizuno T, Takagi T and Nakajima N 1992 Design and fabrication of light driven micropump, *Proc. IEEE Micro Electro Mechanical Systems '92* 31–36

Nelson D F, Wecht K W and Klienman D A 1976 Photophone performance, *J. Acoustic Soc. Am.* **60** 251–255

Rosencwaig A 1980 *Photoacoustics and Photoacoustic Spectroscopy* (New York: Wiley)

Shelley T 1994 Light matched to fast response servo (*Eureka on Campus Spring 1994*)

Snitzer E 1981 'Fly-by-light' sensors, *Aviation Week and Space Technology* 19 January 1981 107

Spooncer R C, Jones B E and Philp G S 1988 Hybrid and resonant sensors and systems with optical fibre links, *J. IERE* **58** (Supplement) S85–S91

Tabib-Azar M and Leane J S 1990 Direct optical control for a silicon microactuator, *Sensors and Actuators* **A21–A23** 229–235

Taft C K and Herrick B M 1981 A proportional piezoelectric electrofluidic pneumatic valve design, *Trans. ASME J. Dyn. Syst. Meas. and Control* **103** 361–365

Thornton K E B, Uttamchandani D and Culshaw B 1986 Temperature dependence of resonant frequency in optically excited diaphragms, *Electronics Lett.* **22** 1232–1234

Tippetts T B and Chapin D W 1976 Tristable fluidic device US Patent 3993101

Uttamchandani D, Thornton K E B and Culshaw B 1987 Optically excited resonant beam pressure sensor, *Electronics Lett.* **23** 1333–1334

Walker A 1990 Optical computing: light at the end of the tunnel? *Physics World* April 1990 25–26

Watts S 1990 Shrinking polymer hides from the light, *New Scientist* 18 August 1990 28

Zdeblick M J and Angell J B 1987 A microminiature electric-to-fluidic valve, *Transducers '87* 827–829

6

Miniature Nanoactuators

Stuart T Smith

This chapter reviews a variety of actuators relevant to fine instrumentation and miniature mechanisms. Within this context the relative merits and limitations of piezoelectric, electrostrictive, electrostatic, electromagnetic, magnetostrictive and shape memory alloys are discussed. These have been chosen because they offer direct electromechanical actuation, are relatively cheap and vacuum compatible.

6.1 INTRODUCTION

Design is a creative process. It usually starts with an innovation of abstract, ideal solutions to a problem and these are then whittled down and refined through knowledge of existing and possible technologies. Although this requires a broad knowledge base, there are ranges of devices that may reasonably be categorized and collected to provide relevant information for the selection stage of the design process. It is the intention of this chapter to briefly review some of the commercially available actuators for achieving controllable displacements with nanometre and subnanometre resolution. At the outset, it is necessary to reduce the scope of this chapter by specifying the range of displacement and scope of applications envisaged as those commonly relevant in fine instrument design. Within this category piezoelectric, electrostrictive, electrostatic, electromagnetic, magnetostrictive and shape memory alloys will be discussed. All of these actuation techniques can be made extremely small, do not require mechanical levers, are vacuum compatible, have good dynamic response and are relatively cheap to implement. Outside this category we are free to choose from a very wide variety of design techniques including friction drives, screwthreads, mechanical levers, hydraulic, Poissons' ratio drives and many others, some of which are more fully described in the book by Smith and Chetwynd (1992). The objective of this chapter is to provide designers with a feel for the individual characteristics of each technique to enable appropriate selection for a specific application.

Before discussing individual actuator mechanisms, a review of common performance measures is presented to clarify to the reader the context in which different terms are used and how these might be applied to design. The reasons for this are twofold. Firstly, it is common that different words may be confused for the same thing or, by omission of relevant information, true performance may be obscured. For example, a device is often specified by its linearity without reference to its repeatability. When choosing an actuator for precision applications, a device based on electrostatic forces may be rejected because of its poor linearity in favour of a piezoelectric device. This could prove to be disastrously erroneous due to the fact that the relationship between electrostatic force and field is, under many circumstances, a lot more precise than that of piezoelectric strain and applied field, which often exhibits considerable hysteresis. Secondly, the performance of the actuator is sometimes hidden by the performance specification of the device within which it forms an integral component. This is illustrated by a loudspeaker in which the stiffness of the actuator might be specified by that of the diaphragm.

The first section reviews those measures commonly used to specify actuator performance and discusses their relevance to fine instrument design. Although by no means a comprehensive overview, the intention of this section is both to refresh the reader of the relevance of various terms and also to provide a checklist for some of the key characteristics that are sometimes absent from a manufacturer's specification. There is then a discussion of individual actuator performance in these terms. Finally, each technique is summarized to provide a quick reference to assist in the earlier, less detailed stages of a design.

6.2 ACTUATOR PERFORMANCE MEASURES

A reason for the selection of one actuator in preference to another is that it more adequately satisfies the desired design criterion. Relevant characteristics encompassing all possible situations are not possible, but there are a number of requirements that are common to many actuator specifications, these being repeatability, linearity, precision, accuracy, load-bearing capability, stiffness, scaling, speed and cost. Assigning priority to these categories is the task of the designer.

6.2.1 Repeatability

As implied by the title, this is a measure of the accuracy with which the position is repeated for a specified demand. Generally, a worst-case scenario is quoted and this is obtained by operating the device over its full-scale deflection, and measuring the largest deviation about a given input demand. This definition is a rather conservative generalization and represents the worst-case performance. Under a wide variety of circumstances, this may be considerably improved by appropriate control or characterization. The most common source of poor repeatability arises through hysteresis in the actuator. This will take on a variety of forms of which by far the most difficult to live with is that which introduces a deadband. This is illustrated in figure 1(a). Here, there is a complete lack of motion during a reversal in demand. Commonly occurring in mechanical couplings, this has serious consequences in that stiffness of the coupling is lost during this transition and high inertial forces are likely to be encountered upon rapid changes in acceleration about the discontinuity resulting in a general lack of controllability. More amenable for control purposes is continuous hysteresis common to field-sensitive actuators (figure 1(b)). Although the open loop control of this type of actuator is restricted by repeatability, this can be overcome through the use of closed loop techniques that are robust enough to be unaffected by the changes in gain. However, the magnitude of this effect will ultimately impose restrictions on dynamic performance and, possibly, precision.

6.2.2 Linearity

A common definition is that a system may be considered linear if the principle of superposition can be applied. In general, this should also apply

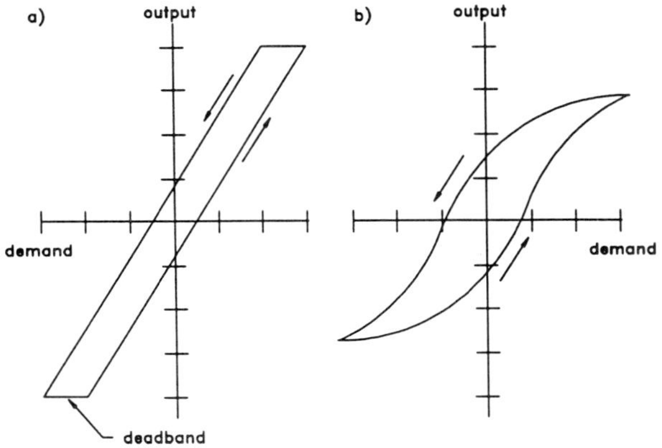

Figure 1. Hysteresis in displacement actuators. a) deadband and b) continuous.

to linear dynamic systems in which transient and steady state responses can be combined to give a reasonable estimate of displacement (in our case). Under 'static' conditions the relationship between displacement and demand should be a straight line. This can be deceptive as a monitor of performance as some actuators may not be linear, although the non-linearity may be very reproducible. Under some circumstances, discrepancies between theory and experiment are, as yet, unmeasurable as for the case of force between two electrodes in a vacuum which exhibits a square-law dependence with potential difference. Using modern computer hardware and analogue electronics, it is often possible to introduce 'hidden' signal conditioning algorithms or elements such that the actuator appears linear. Under these circumstances, it is therefore more appropriate to concern ourselves with repeatability and, mainly for this reason, it is the author's view that this parameter is of little importance to modern instrument designers.

6.2.3 Precision

Precision is often referred to as the ratio of range to resolution. However, it should not be thought of as a parameter that remains of equal significance regardless of scale. As demand increases for instruments and machines of higher resolution, the designer is moving further towards unknown technologies and design techniques and correspondingly towards increasing cost. A recent breakthrough in high-resolution mechanical design has been seen with the development of probe microscopy techniques. Initially invented in the early 1970s by Young (1972), it was not until 1982 that atomic resolution and, indeed reasonable images of nanometre dimension were obtained by Binnig *et al.* It is only recently that submillimetre sized images are being achieved (Fu *et al* 1992), although precisions of one part in 10^5 have been commonplace in conventional 'precision engineering' for many decades.

6.2.4 Accuracy

This is the nearness of the measured value to the 'true' value which is, in turn, a value commonly agreed in the scientific community. Not being absolutely realizable, this is usually specified as a tolerance which must be directly traceable to current standards. This is clearly independent of precision which does not rely upon calibration. An instrument will retain its precision no matter how poorly it is calibrated and this is often a source of confusion that occurs frequently in performance specification of instrumentation.

6.2.5 Load-bearing Capability

By definition, actuators are required to do work often requiring the movement of a mass against a gravitational field. A primary requirement is that the system does not change its operating characteristics under these conditions.

6.2.6 Stiffness

In some ways, the stiffness is often functionally related to the previous category. However, it is necessary to discriminate between two type of actuator. Field-coupled actuators utilize gradients in potential to generate forces which are then translated into motion through a mechanical spring system. For this type of device, the apparent stiffness of the actuator becomes that of the spring combined with that of the actuator. Under certain conditions, the stiffness of the actuator can be made extremely small and is therefore virtually decoupled from the system. Those actuators based on induced strain within a material upon direct application of an electric field tend to have a stiffness that is a function of the material properties and geometry. Consequently, there is a wide variety of stiffnesses available from actuators, and care must be taken in their specification.

6.2.7 Scaling

Microelectro-mechanical systems have recently emerged as a fascinating and exciting approach to system design and miniaturization. There is great interest in the ability to produce smaller and smaller instruments and machines. The effect of scaling is to reduce the power requirements of instrumentation as well as increasing the speed of response and resistance to body forces such as self-weight. As these devices become smaller and smaller, so too must the actuators.

6.2.8 Speed

Dynamic response will always be a challenge to instrument designers. This tends to be related to stiffness and ability to scale.

6.2.9 Cost

In attempting to assess costs, a number of assumptions and some intuitive guesswork have been used. For each actuator, the author has ascertained the cost of materials, machining of auxiliary components and necessary electronics. A relative scale has then been created. Consequently, such costs as development, marketing and maintenance are not included and this must therefore be considered as only a rough ratiometric scaling.

6.3 PIEZOELECTRIC DRIVES

Of the twenty-one crystal classes that do not have a centre of symmetry, twenty of these can experience a dimensional change upon application of an electrical potential gradient (electric field) and such materials are known as piezoelectric. This lack of a centre of symmetry is a necessary precursor for this phenomenon. Ten of these classes will also generate a surface charge upon heating and these are known as pyroelectric materials. Although *all pyroelectrics are piezoelectric*, the reverse is not always true. This is due to the fundamentally different reasons for the occurrence of these two effects. In a pyroelectric material, the net dipole is changed by a change in the magnitude of polarization when heated. In piezoelectric materials that are not pyroelectric, the dipoles which are invariably altered in the presence of heat are so arranged that they compensate each other resulting in no net change in dipole moment. Upon application of a stress, some dipole moments in specific directions are increased more than the others and a net dipole change is induced. Additionally, some materials can be polarized by the application of a sufficiently strong electric field. By analogy with ferromagnetism, these are called ferroelectric although neither necessarily contain iron. *All ferroelectrics are both piezoelectric and pyroelectric* and constitute all of the commonly used actuator materials because, like magnetic materials, they can be easily and permanently polarized (commonly referred to as poling) at, or near to, their Curie temperatures. This enables these materials first to be sintered as a ceramic and then polarized in a desired direction. Poling, however, does not result in perfect polarization in these polycrystalline materials but results in the creation of a large number of domains in which there is an overall polarization direction. If one imagines an ideal material in which there are N parallel molecules of length δ having a charge of $+q$ and $-q$ at each end, then the magnitude of the total polarization is given by

$$P_x = Nq\delta_x. \tag{1}$$

The only way to change this value is to physically squeeze the poles together by the application of an applied stress, or to displace the charges with an applied field. Under uniaxial conditions, the total displacement current, D, is given as a linear combination of the two phenomena i.e.

$$D_x = \varepsilon E_x + dT_x \tag{2}$$

where ε is the materials permittivity at constant stress (F m^{-1}), d is the piezoelectric strain constant (C N^{-1} or m V^{-1}), T is the applied normal stress (N m^{-2}) and E is the field strength (V m^{-1}). In fact, this treatment is very much an oversimplification and in practice these effects are heavily anisotropic and a more complete analysis requires tensor analysis. Excellent introductory reviews can be found in the papers of Jaffe and Berlincourt

(1965) and Gallego-Juarez (1989). Conversely, the strain s for a given applied field or stress is given by the equation

$$S_x = d_t E_x + sT_x \tag{3}$$

where d_t is the transposed piezoelectric strain matrix (m V^{-1}) and s is the elastic compliance at constant applied field (m^2 N^{-1}).

Again values for these constants and other interrelated properties for a wide range of materials can be found in the paper of Jaffe and Berlincourt (1965). It should be stressed that not only are most of these materials pyroelectric and thus sensitive to temperature, but there is an applied field above which the dielectric strength diminishes catastrophically. For the very popular lead zirconate titanate (PZT) material at 25 °C this will be in the region 0.7–1.0 kV mm^{-1}.

From the above simple analysis it can be seen that, in the absence of an applied stress, there will be a net displaced current required to obtain a proportionate strain. It is usual to use voltage drives for this type of actuator and, as such, the actual displacement will depend upon the size, distribution and stability of the electric domains within these polycrystalline ceramics as well as other influences such as homogeneity of the material. Because of these faults, the resulting relationship between applied voltage and displacement usually exhibits considerable hysteresis, usually in the region of 15%. The displacement for an applied potential difference (typically in the region of −1000 V for micrometre expansions) across a typical piezoelectric actuator is shown in figure 2. Methods for the reduction of this effect using feedback control, computer linearization techniques or a charge drive have been variously applied. The latter approach was first proposed by

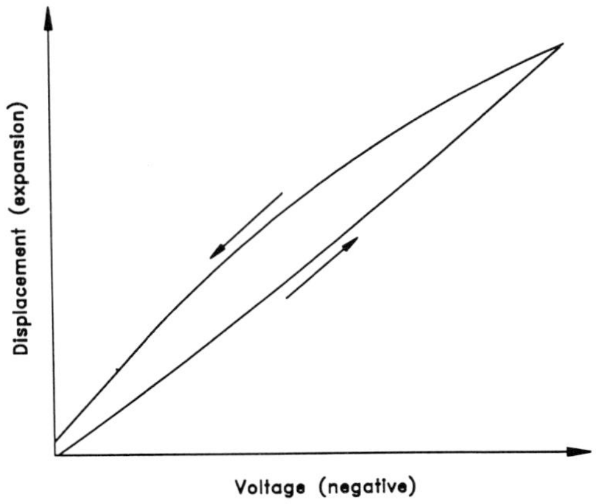

Figure 2. Displacement–voltage characteristic of a typical piezoelectric actuator.

Newcomb and Flinn (1982), who suggested using a charge amplifier as the driving element. This was further simplified by Kaizuka and Siu (1988) and Kaizuka (1989), by inserting a compensating capacitor in series with the piezoelectric and using a simple voltage drive. Using this technique the percentage hysteresis could be reduced by up to one fifth with, however, a similar reduction in the sensitivity. More importantly, creep is also reduced in similar proportion. Creep is caused by the gradual changing of domains and will quite commonly be in the region of 30% over a period of a few to tens of minutes (Jaffe *et al* 1971), which precludes these devices for metrological application in open loop operation (Drake *et al* 1986, van de Leemput *et al* 1991). The magnitude of these effects is very difficult to assess and control and will vary from material to material and even from batch to batch. Another disadvantage with these materials is the cross coupling which not only results in a lateral expansion of approximate magnitude 0.25–0.65 but often leads to a bending motion that can result in angular deflections that can be a significant part of the motion in miniature actuators. It is more often less than a few percent of the total motion and some modern tubular actuators now produced may have parasitic tilts of less than one arc second per micrometre of traverse.

To overcome the non-linearities associated with this actuator, closed loop control strategies have been developed and devices are now commercially available. Using capacitive feedback techniques, Hicks *et al* (1984) were able to control Fabry-Perot etalons (parallel optical flats for use in spectrometry) at a dynamic gap separation maintained constant at the 1 pm $Hz^{-1/2}$ level and commercial devices are now available which operate a range/resolution ratio of 10 000:1. The feedback transducers for these applications include capacitance displacement sensors and direct strain gauging. Generally, a displacement of approximately 10 μm is obtained at an applied field of 1000 V. Although it is possible to increase in scale, as the devices become larger so too does the voltage necessary to obtain the same field strengths. Consequently, they are more commonly applied to small-scale manipulators of the order of tens of millimetres or less in dimension. One of the key features of such devices is their speed of response which is limited by the speed of sound in the material and is inversely proportional to length. For a PZT device, the speed of sound in the material along the polar axis is sufficiently high to enable operation at gigahertz frequencies, although for large displacements hysteresis and other losses will require the dissipation of heat. There has recently been a surge in the use of these devices for scanning tunnelling microscopes where the tip is controlled by three orthogonal piezoelectric actuators which can collectively occupy less than two cubic millimetres (Smith and Binnig 1986).

The apparent limitations on the available displacement are overcome by either stacking many actuators in series, or creating a bi- or multi-laminar element analogous to the bimetallic strip (van Randeraat and Setterington

1974). Three such arrangements are shown in figure 3. As the number of actuators increases, both the stiffness and the dynamic performance of these devices will reduce. The bimorph of figure 3(b) can be further enhanced by arranging the poling such that the actuator naturally bends in an 'S' shape. This relieves the flexures from restriction due to the end clamps. Using such a method Matey *et al* (1987), were able to obtain displacements of 120 μm in all three axes with the resolution limited by electronic and mechanical noise. There is evidence that the hysteresis will increase due to creep in the bonds. This effect, which theoretically reduces with temperature, has been observed by Blackford *et al* (1987), and Tritt *et al* (1987). The lowest natural frequency of the displacement stage presented in these papers was 190 Hz.

Finally, for even longer-range applications a device called an 'inchworm' has been developed. As the name suggests, this motion is achieved in a manner similar to that used by some caterpillars. A piezoelectric actuator is mounted onto either a slideway or an optical flat, having an electrically conducting surface. Attached to the base of the actuator are two or more conducting pads separated from contact with the slideway by a thin layer of insulating material. A piezoelectric ceramic is then used as the body connecting the pads. The inchworm is actuated by first electrostatically clamping one pad to the sideway, and then extending the piezoelectric material. The free end is then clamped and the fixed end is released. By then contracting the actuator and repeating these steps, it is possible to obtain unlimited motion. This type of device was first proposed by Joyce and Wilson (1969), became famous as the coarse feed in the first scanning

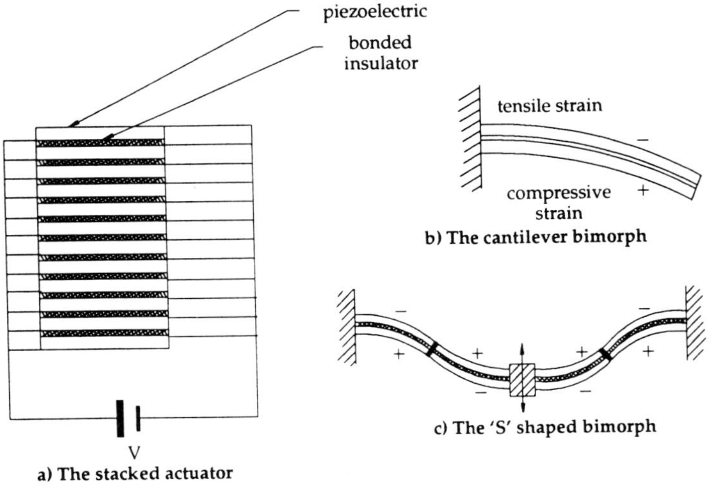

Figure 3. Long-range piezoelectric actuators.

tunnelling microscope of Binnig and Rohrer (1982), and was subsequently investigated by many commercial organizations. Although problems with stability of these actuators have prevented their being commercially available until around 1989 (Mamin *et al* 1985), devices are now available with a traverse of up to 100 mm at individual step sizes of 4 nm or less and traverse rates of up to 2 mm s^{-1}. A less accurate but still interesting extension of this idea is the piezoelectric motor that operates on similar principles (there are now versions that operate by creating a circulating ripple (Tojo and Sugihara 1985, Hatsuzawa *et al* 1986)).

6.4 ELECTROSTRICTIVE DRIVES

Although electrostrictive ceramics were first discovered in 1954, they have only recently become commercially available (Isupov *et al* 1989). A combination of lead magnesium niobate and lead scandium niobate is now known to exhibit a giant electrostrictive effect giving an extension of approximately 0.7 nm V^{-1}. More commonly used is a ceramic made up from 90% lead magnesium niobate and 10% lead titanate (Uchino *et al* 1980, 1981). This phenomenon is similar in nature to the piezoelectric effect, only in this instance strain can be conceptualized as the deformation due to a field gradient, but not a direct polar strain. In this instance, individual microdomains can be considered independent in their action. Polar strain is not the primary mechanism responsible for distortion, and electrostriction may be better considered at this moment in time to be independent of the piezoelectric effect. Simplistically, it may be envisaged as the result of rotation of the polar domains within the ceramic. The ceramic is actuated by applying a field transverse to the polar direction inducing rotation of the domains resulting in behaviour characteristic of electrostatic forces in that the extension as a function of the applied field (V m^{-1}) appears to follow a square law at lower values. Because of the independence of individual domains there is no need to polarize this before use. This then becomes more linear until saturation is reached at around 10 kV cm^{-1} (Uchino *et al* 1981). Interestingly, due to similarities in the actuation mechanism, the strain/applied field shape characteristic for this material is of a similar form to magnetostrictive materials.

The properties of this material for use as a displacement actuator are remarkably similar to those of piezoelectric materials (Eyraud *et al* 1988). Most researchers operate these at a bias to avoid operating in the non-linear region and, in particular, because of the low sensitivity at small potential. Analogous to a typical piezoelectric actuator, displacements of around 10 μm are achievable at applied fields in the region 0–1 kV operating at frequencies up to 20 kHz or more. Recently, stacked systems have been produced that

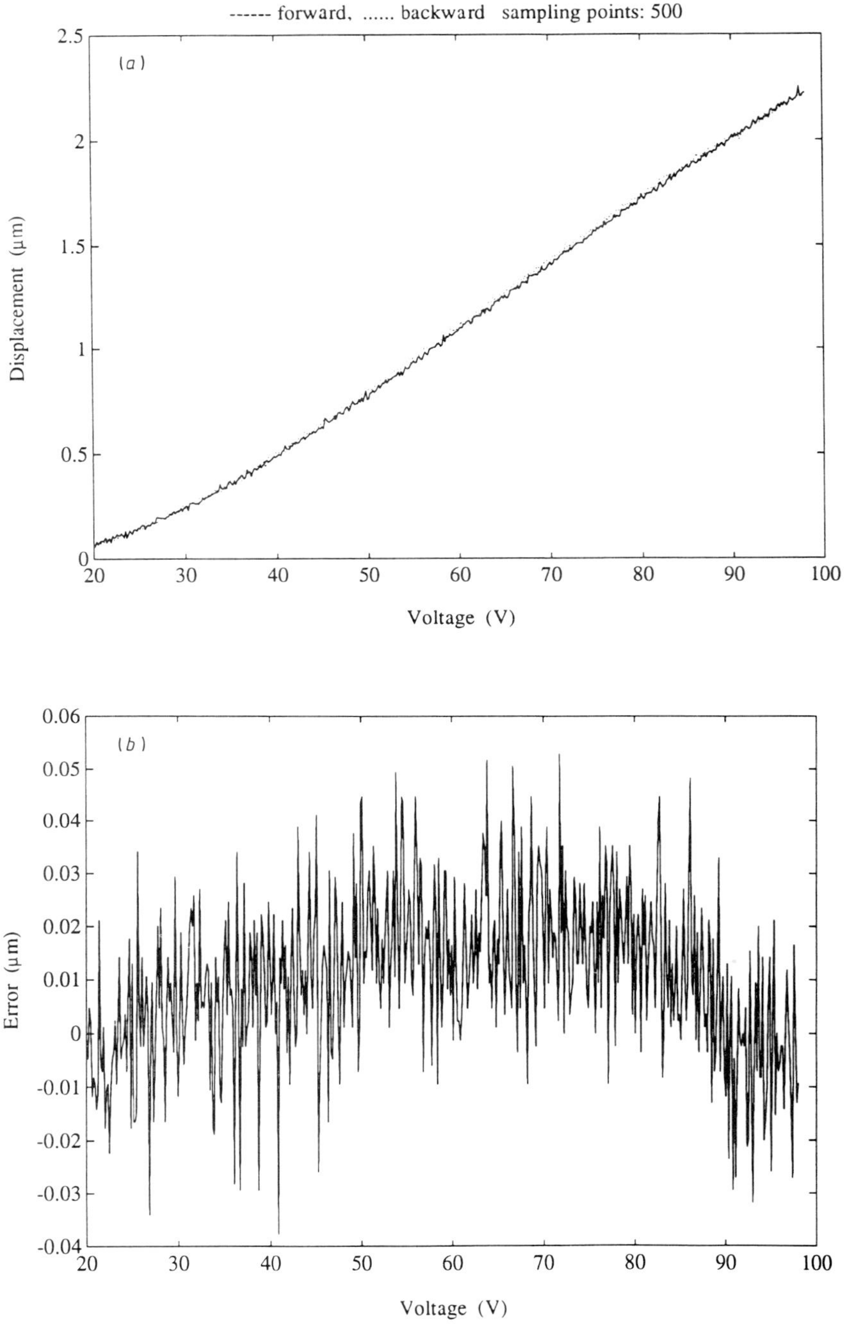

Figure 4. Displacement–voltage characteristic of a low-voltage electrostrictive actuator. (*a*) displacement against applied voltage in forward and reverse traverse and (*b*) error between the two directions.

can achieve displacements of up to 15 μm with applied voltages around 150 V. Figure 4(a) shows a typical forward and reverse expansion of a stacked system of diameter 5 mm and length approximately 10 mm with a ramp of 80 V and a 20 V bias. The important difference between electrostrictive and piezoelectric actuators is that the former has hysteresis in the region of a few percent or less at room temperature, and a relatively low expansion coefficient in the region 10^{-6} K^{-1}. Again, there is much ambiguity with the former value because of a general confusion between linearity and hysteresis. Subtracting the two traces of figure 4(a) results in the error plot of figure 4(b). This clearly indicates that for a displacement range of just over 2 μm there is a difference between extension and contraction at a given voltage of less than 20 nm which is approaching the resolution of the stylus measurement system used in these tests. The expansion coefficient is similar in magnitude to some of the best ceramic materials making this ideal for precision actuator applications where the generation of heat may introduce undesirable errors. Reducing the temperature of the actuator below 270 K will result in a rapid increase in hysteresis obviating its benefit over the currently cheaper piezoelectric ceramics.

6.5 ELECTROMAGNETIC DRIVES

There is a large number of possibilities for the design of electromagnetic actuators. Most, however, involve the use of the magnetization and demagnetization of soft magnets. To restrict the range of devices considered in this section, only those linear mechanisms utilizing permanent magnets in the presence of magnetic fields insufficient to induce significant demagnetization will be considered. The most common AC linear displacement actuator is the audio loud speaker. In precision electromagnetic devices, a saturated permanent magnet (usually SaCo or NdBFe) is surrounded by a coil (or vice versa) and a force proportional to both the strength of the magnet and the gradient of the field perpendicular to the axis of its poles (and thus coil current) is generated between the two. To obtain a controlled displacement, the force generated is usually applied to a linear spring and it is this which appears to be the main limitation to the achievable accuracy of such a device. For a saturated magnet in a relatively weak field, the force/current characteristic has been found to be linear to within better than one part in 10 000. Using a spring manufactured from single-crystal silicon it is possible to obtain open loop displacements of up to 100 nm with a resolution of 5 pm as measured using x-ray interferometry. This type of actuator requires some form of stiffness against which to act and can therefore only be considered as a primary actuator in need of some secondary mechanism by which controlled displacements may be achieved. Because of this the available range and resolution are not easy to assess. In the paper of Smith and Chetwynd (1990), it is shown that the output force per unit input power is dependent upon both the permanent magnet material (in particular its intrinsic magnetization) and the conductivity of the input windings. Clearly the use of superconductors will relieve these limitations and if achievable at reasonably high temperature would enable the implementation of extremely linear, lower-power actuators. The linearity of present devices is proportional to the fractional demagnetization and this is, in turn, closely related to the recoil permeability which should be as low as possible for maximum linearity. Demagnetization curves for the materials NdDyFeBNb, NdDyFeB and NdFeB are shown in figure 5 (reproduced from Parker *et al* 1987). Non-linearities are due to demagnetization as the intrinsic magnetization follows the characteristic curve, and complete demagnetization will occur in the presence of sufficiently high field strengths. Conversely, the recoil permeability will also be related to the applied field and reduces to zero at very low fields. In normal applications, relatively weak fields are employed and it is the constancy of the *B–H* characteristic that limits the linearity of operation. From figure 5, it can be seen that the simple NdFeB material has the more nearly constant characteristic in the presence of a low field strength with the alloys of dysprosium and niobium creating a larger slope, but also considerably increasing the coercive strength. However, the author has not been able to obtain characteristics with precision of better than 1 part in 10^3. The Curie temperature (the temperature at which permanent magnetization is lost) is approximately 270 °C for these materials although it can be increased to approximately 500 °C by the addition of cobalt (Sagawa *et al* 1984).

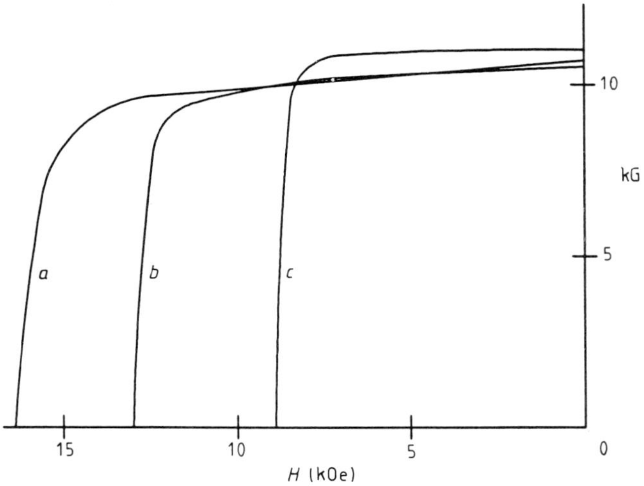

Figure 5. Demagnetization curves for NdBFe magnet materials. a) NdFeDyBNb, b) NdFeDyB and c) NdFeB.

The two different ways of utilizing an electromagnetic force actuator are either to drive the coil or the magnet. In the first case, a permanent magnet is constructed to consist of a permeable loop that is broken by a small gap in which the driven coil of winding length L is positioned. By passing a current I through this coil it will experience an axial force, F_x, given by (for a detailed discussion see Hadfield (1962))

$$F_x = B_x IL \tag{4}$$

where B_x is the flux density across the gap and from considerations of continuity this can be approximated from the equation

$$B_x = \sqrt{\frac{\mu_0 H_{\mathrm{m}} B_{\mathrm{m}} V_{\mathrm{m}}}{V_{\mathrm{g}}}} \tag{5}$$

where μ_0 is the permeability of the gap, $H_{\mathrm{m}}B_{\mathrm{m}}$ is the energy product for the magnetic material and V_{m} and V_{g} are the volumes of the magnet and gap respectively.

From equation (5), it can be seen that the optimum condition is obtained for a magnetic material having the highest energy product and the minimum gap volume. Thus, if there is no restriction on space, it is best to use the largest volume of magnet possible. The biggest drawback with this system is the fact that the active element, to which power must be supplied, is attached to the moving stage, and energization of the winding will create a source of heat in the mechanism. One way of overcoming this problem is to use two counter rotating windings and operate the actuator in differential mode while maintaining a constant input power. This should, in practice, result in a constant heat input that will eventually reach an equilibrium temperature and remain constant during operation. The disadvantage with this method is that the extra winding is redundant in terms of the actuating force, requiring additional gap space and therefore reducing the force per

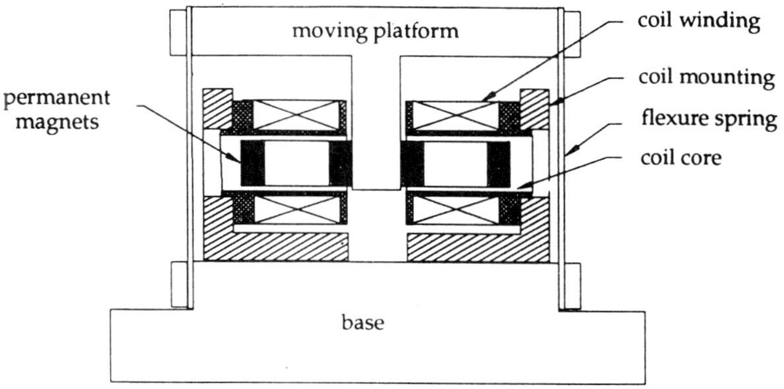

Figure 6. A two-coil, four-magnet, solenoid–magnet-driven linear spring.

unit power and adding mass to the moving stage. Because of the large mass of magnet required for this type of actuator, it is not usual to attach the passive magnets to the moving platform. To utilize such a device, a different principle of operation must be employed. In this, the permanent magnet is attached to the moving platform and this is, in turn, surrounded by a coil that is rigidly mounted onto the base of the device. The force on the magnet is then proportional to both the magnitude of the magnetic moment and the gradient of the field along its axis of magnetization. In the case of a permanent magnetic material having a constant flux density with applied field (a condition nearly satisfied by the selection of a magnet of low recoil permeability and avoiding strong fields as mentioned above) the moment is equal to the intrinsic magnetization multiplied by its volume. Additionally, under these conditions, the intrinsic magnetization may reasonably be equated to the remanence of the magnet, B_{rem}, and the resultant force on a magnet of volume V_m is

$$F_x = B_{rem} V_m \frac{H_{x1} - H_{x2}}{x1 - x2} \tag{6}$$

where H_{x1} and H_{x2} are the values of applied field at each of the poles of the magnet which are situated at positions $x1$ and $x2$ (these are usually near to the end of a permanent magnet) and F_x is the force along the x axis which is colinear with the vector $(x1, x2)$. For a relatively short magnet in a field of gentle gradient this can be written in differential form

$$F_x = B_{rem} V_m \frac{dH_x}{dx} \tag{7}$$

A simple actuator can be constructed by fixing or glueing a series of disc-type magnets to a linear spring device, and surrounding these by a uniformly wound circular cylindrical coil. The maximum field gradient in this instance is near to each end of the coil and thus it is favourable to place a magnet at this position with opposing poles (the gradients are equal and opposite at each end), see figure 6. This type of device has the advantages that the force on the magnet is relatively insensitive to position and there is no mechanical coupling which results in a near-zero stiffness force actuator. A disadvantage is that the coil should be as close to the magnets as possible. This is hard to achieve without the magnets touching the winding spool. A useful method for checking this is to electrically isolate the magnets from the spool and then check for a short circuit as an indicator of mechanical contact. The optimum geometry of the coil for such an actuator is presented in the paper of Smith and Chetwynd (1990), while the design parameters for a coil having a uniform field gradient are presented in the paper of Pouladian-Kari *et al* (1988).

At present, it is possible that the common loud speaker may be one of the most efficient mechanisms achieving displacements in the region of 500 μm over a 20 kHz bandwidth. However, the flexure would have to be considerably redesigned if dimensional stability is to be achieved. Given this restriction it is feasible that nanometre-level resolution may ultimately be possible.

6.6 ELECTROSTATIC DRIVES

As with the above actuator, this is a non-contact-force-driven device, utilizing the principle that any two electrodes separated by an insulating barrier will be attracted towards one another if there is an electric potential difference between them. The magnitude of this force for a rectangular, parallel plate capacitor overhanging the edge of a flat electrode is given by

$$\begin{aligned} F_a &= \frac{\varepsilon b V^2}{2x} \\ F_x &= \frac{\varepsilon a b V^2}{2x^2} \end{aligned} \qquad (8)$$

where a is the length of the overlap of the two electrodes, b is the width, and x is the perpendicular separation.

From this it can be seen that the force is always attractive and is related to the permittivity (ε), the reciprocal of the separation and also the square of the potential. This force can then be changed by varying either the polarization or the applied potential to effect a change in the separation. The first equation indicates that the force F_a acting parallel to the plane of the electrodes is independent of the overlap dimension. It is for this reason that this type of actuator is only used with extremely small separations as is the case for micromechanical devices (Trimmer and Gabriel 1987). Stiffness is related to both the square of the ratio of voltage to stiffness and the dielectric constant. For an area of 100 mm^2, a separation of 0.01 mm and a bias voltage of 100 V, such a device operating in air would have a stiffness of magnitude 4.3×10^{-4} N mm^{-1} and will generate a force of 4.3×10^{-8} N. Thus because of the very low forces associated with this type of actuator it is most commonly used for ultrasonic excitation, microminiature devices such as submillimetre-sized motors (Fan *et al* 1989, Tai and Muller 1989), microelectronic resonance accelerometers and may find application in resonant probe microscopy for the combined excitation and monitoring of small probes. Although the force characteristic is non-linear, it is an exact relationship extending over many orders of magnitude and restricted mainly by the dielectric value of the insulating barrier. A disadvantage of this type of device is that if air is used as the separating dielectric, then it will vary with both temperature and humidity and thus for precise applications a closely controlled environment is required.

6.7 MAGNETOSTRICTION

This phenomenon is very similar to that of the piezoelectric effect only it involves a change in dimension in the presence of a *uniform* magnetic field. This means that the displacement per unit field will increase with dimension making this favourable as a large-scale or heavy-duty actuator. As would be expected, this phenomenon also has intrinsic hysteresis and will creep with time. A material demanding recent interest is an alloy of terbium, dysprosium and iron (called Terfenol-D) which has giant magnetostrictive properties.

Figure 7 is a plot of strain against magnetic field. When the material is subject to a stress of $\simeq 20$ MPa, a moderate field strength of 70 kA m^{-1} will result in a strain of 700×10^{-6} (reproduced from Clark *et al* 1988). From this it can be seen that the strain is also dependent upon the applied stress and will fall off rapidly either side of this value. Consequently, this type of actuator will require a preload to ensure that, for this material composition, the working stress is maintained between 8 and 20 MPa (Oswin *et al* 1988). The size of such an actuator is restricted primarily by the cost of the

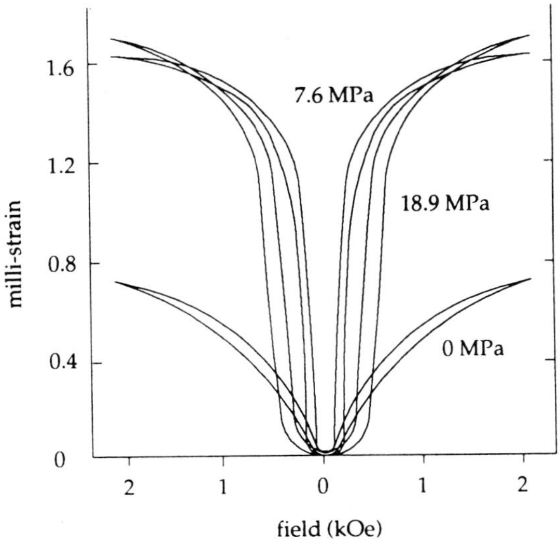

Figure 7. Strain-magnetic field characteristic of Terfenol-D under different applied stresses.

material (terbium is very expensive at this moment in time) and that of the field generator. A reasonable device for the control of a grinding wheel (say) might have dimensions of 100 mm in length and an area of 20 000 mm^2 giving a stiffness of 200E and a displacement of 80 μm under a load of 400 kN. This clearly has potential as a heavy-duty nanoactuator under feedback control. The speed of response of such a device is restricted mainly by the inertia of the material and for the large scale applications being considered, a typical resonant frequency of 1 kHz may be expected. Engdahl and Svensson (1988) and Yuan *et al* (1987) have used this type of actuator for the closed loop control of a tool-feed compensator with a 125 Hz bandwidth at 10 nm resolution. However, as the size, and thus displacement, is reduced, devices having an operating frequency of up to 300 kHz are possible (Oswin *et al* 1988).

6.8 MAGNETOELASTICITY

This is frequently referred to as the ΔE effect and is related to the magnetostrictive coefficient (Squire 1990). All magnetostrictive devices will change elastic modulus in the presence of a uniform magnetic field and it is this, in part, that contributes to the giant magnetostrictive effect. To utilize this as a mechanical actuator, the material must first be subjected to strain via a dead weight, spring or internal stresses. These will result in a distortion of the material, the magnitude of which depends on its elastic modulus. If the material is then placed within a magnetic field, the elastic modulus and thus the distortion will change. The magnitude of this distortion is dependent upon both the percentage change in elastic modulus and the magnitude of the initial distortion. A typical value for percentage change is around 2–10% and the maximum strain for many materials will be in the region of 0.1%. Thus the achievable strain is approximately 10–50 microstrain or a change in length of 2–10 μm for a 100 mm long device. A material of particular interest is Elinvar or Ni-Span-C which is a 42% Ni–Fe alloy. This composition results in a near-zero thermoelastic coefficient near to room temperature and a thermal expansion coefficient of 3×10^{-7} K^{-1}. It does, however, undergo a change in Young's modulus of approximately 0.1% in the presence of a high magnetic field. Thus it is feasible to use this as a magnetoelastic device that is dimensionally sensitive to magnetic field but is relatively immune to small temperature variations. The main advantage of this type of actuator over all others is that the displacement per unit field strength will depend upon the load condition and therefore will be *tuneable*. Although this is a rather small effect, there are materials with much greater sensitivities to the electric field (in particular Terfenol-D, and up to 20% strains are possible with some Ni–Fe and Co–Fe alloys (Bates 1953)).

Additionally, the change in elastic modulus is not linear and can have both positive and negative slope, and can contain considerable hysteresis making this more suitable as part of a closed loop system. Again, the stiffness and the propensity to large-scale devices for such an actuator is similar to the above only the cost of alloys is a lot less for this actuator.

6.9 SHAPE MEMORY ALLOYS AND BIMETALLIC STRIPS

This family of devices requires heat as the actuating mechanism. In shape memory alloys (usually Ni–Ti) the material is bent at a high temperature and then allowed to cool, whereby it returns to its original shape. Upon reheating the material will return to its deformed shape at that original temperature, as if it had remembered. For a review of these materials see Schetky (1979), or Golestaneh (1984). Actuators using this effect are capable of very large displacements for a relatively small device size. A typical example would be a deflection of 20 mm for a cantilever beam of around 100 mm in length and a drive force of 300 N, with a time constant of approximately 4 seconds (Yaeger 1984). Heating is usually applied by passing a current through the actuator. More recently, smaller actuators have been used to align optical fibres by providing the pushing action to position them in precisely formed vee grooves. Using this technique, it was possible to position an individual fibre to within 100 nm (Jebens *et al* 1989). There is also interest in using this type of actuator for miniature robot arms (Kuribayashi 1989).

Bimetallic strips perform similarly, the main difference being that the mode of operation is differential thermal expansion of two intimately contacted thin materials and is therefore a function of temperature. A typical example is the old vehicle indicator mechanism. Again, under controlled conditions, it is likely that similar range and resolution to the shape memory alloys may be achieved.

6.10 SUMMARY

This short chapter has reviewed some of the actuator mechanisms that may be utilized for small-scale translation devices with nanometre-level capability. As a quick aide-mémoire, some of the major advantages and limitations are outlined below.

Piezoelectric actuators can provide stiff, load-bearing, miniature actuators with low thermal expansion and a high frequency response. Ideally suited to small scale actuators, stacked systems can be produced to provide ranges of many hundreds of micrometres without resort to mechanical lever systems. Disadvantages are hysteresis and creep, both of which are of the order 15–20% and applied voltages in the region of 1–2 kV.

Electrostrictive-based actuators offer similar load-bearing and dynamic performance to their piezoelectric counterpart, with the advantages of low hysteresis and a low voltage operation resulting in reduced cost and enhanced controllability. Although materials have been known to exist for many years, it is a relative newcomer to the actuator market.

Electrostatic This phenomenon is used to create a known force which must then be applied to a spring system. Use of this in a known environment or vacuum will result in forces that can be calculated with extremely high accuracy, and therefore it offers potentially the most precise translation mechanism. Unless small separations can be obtained, the magnitude of readily obtainable forces is low. Therefore it is most suited to small scale actuators. Requiring only that the two adjacent surfaces are electrical conductors and a voltage supply, these are likely to constitute a low-cost actuator.

Magnetostriction Conversely, this requires a uniform field and displacement range scales linearly with dimension, making it ideal as a large-scale, high-force and stiffness actuator. Relying upon magnetic induction, it exhibits considerable hysteresis. The high cost of terbium and field generating electrics has restricted this actuator to specialist applications.

Magnetoelasticity offers similar performance to the above with the fundamental difference that the magnitude of displacement relies linearly on the initial stress. Consequently, it should in principle be possible, through design, to tune the displacement range.

Shape memory alloys and bimetallic strips These will form low stiffness but large range actuators. Applications have been mainly in micromechanical robot arms or large conventional engineering mechanisms. Because of their slow response they are mainly operated in a pseudo on–off mode and constitute a heat source which will result in temperature gradients within the device. The dynamic response of this device is almost invariably very slow. For miniature actuators the cost is likely to be low, requiring the material and a current source.

REFERENCES

Bates L F 1961 *Modern Magnetism* 4th edn (Cambridge: Cambridge University Press) ch XI

Binnig G and Rohrer H 1982 Scanning tunnelling microscopy *Helv. Phys. Acta* **55** 726–735

Blackford B L, Dahn D C and Jericho M H 1987 High-stability bimorph scanning tunnelling microscope *Rev. Sci. Instrum.* **58** 1343–1348

Blackford B L and Jericho M H 1990 Simple two dimensional piezoelectric micropositioner for scanning tunnelling microscope *Rev. Sci. Instrum.* **61** 182–184

Clark A E, Teter J P and McMasters O D 1988 Magnetostriction 'jumps' in twinned $Tb_{0.3}Dy_{0.7}Fe_{1.9}$ *J. Appl. Phys.* **63** 3910–3912

Drake B, Sonnenfeld R, Schneir J, Hansma P K, Slough G and Coleman R V 1986 Tunnelling microscope for operation in air or fluids *Rev. Sci. Instrum.* **53** 441–445

Engdahl G and Svensson L 1988 Simulation of the magnetostrictive performance of Terfenol-D in mechanical devices *J. Appl. Phys.* **63** 3924–3926

Eyraud L, Eyraud P, Gonnard P and Troccaz M 1988 Materiaux electrostrictifs pour actuateurs *Revue Phys. Appl.* **023** 879–889

Fan L, Tai Y and Muller R S 1989 IC-processed electrostatic micromotors *Sensors and Actuators* **20** 41–47

Fu J, Young R D and Vorburger T V 1992 Long-range scanning for scanning tunnelling microscopy *Rev. Sci. Instrum.* **63** 2200–2205

Gallego-Juarez J A 1989 Piezoelectric ceramics and ultrasonic transducers *J. Phys. E: Sci. Instrum.* **22** 804–816

Golestaneh A A 1984 Shape-memory Phenomena *Physics Today* **37** 62–70

Hadfield D 1962 *Permanent Magnets and Magnetism* (London: Wiley) chs 2, 6 & 7

Hatsuzawa H, Toyoda K and Tanimura Y 1986 Speed control characteristics and digital servo system of a circular travelling wave motor *Rev. Sci. Instrum.* **57** 2886–2890

Hicks T R, Reay N K and Atherton P D 1984 The application of capacitance micrometry to the control of Fabry-Perot etalons *J. Phys. E: Sci. Instrum.* **17** 49–55

Isupov V A, Smirnova E P, Yushin N K and Sotnikov A V 1989 Electrostrictor ceramics: Physics, materials applications *Ferroelectrics* **95** 179–183

Jaffe H and Berlincourt D A 1965 Piezoelectric transducer materials *Proc. IEEE* **53** 1372–1386

Jaffe B, Cook W R and Jaffe H 1971 *Piezoelectric Ceramics* (vol 3 Series of Monographs on Non-Metallic Solids) (London: Academic) pp 83–85

Jebens R, Trimmer W and Walker J 1989 Micro actuators for aligning optical fibers *Proc. 2nd IEEE Workshop on Micro-Electro Mechanical Systems* (*Salt Lake City*) 35–39

Joyce G C and Wilson G C 1969 Micro-step motor *J. Phys. E: Sci. Instrum.* **2** 661–663

Kaizuka H and Siu B 1988 A simple way to reduce hysteresis and creep when using piezoelectric actuators *Japan. J. Appl. Phys.* **27** L773–L776

Kaizuka H 1989 Application of capacitor insertion method to scanning tunnelling microscope *Rev. Sci. Instrum.* **60** 3119–3122

Kuribayashi K 1989 Millimetre-sized joint actuator using a shape memory alloy *Sensors and Actuators* **20** 57–64

Mamin H J, Abraham D W, Ganz E and Clarke J 1985 Two-dimensional, remote micropositioner for a scanning tunnelling microscope *Rev. Sci. Instrum.* **56** 2168–2170

Matey J R, Crandall R S and Bryki B 1987 Bimorph-driven *x-y-z* translation stage for scanned image microscopy *Rev. Sci. Instrum.* **58** 567–570

Newcomb C V and Flinn I 1982 Improving the linearity of piezoelectric ceramic actuators *Electronics Lett.* **18** 442–444

Oswin J R, Edenborough R J and Pitman K 1988 Rare-earth magnetostriction and low frequency sonar transducers (*IOP*/IEE/IEEE day seminar on Magnetic Sensors and Amorphous Materials, April 1988) *Aerospace Dynamics* **24** 9–13

Parker S F H, Pollard R J, Lord D G and Grundy P J 1987 Precipitation in NdFeB-type materials *Trans. IEEE* **MAG-23** 2103–2105

Pouladian-Kari R, Parkes D, Jones R M and Benson T M 1988 A multiple coil solenoid to provide an axial magnetic field with a near-linear gradient *J. Phys. E: Sci. Instrum.* **21** 557–559

Sagawa M, Fujimmura S, Yamamoto H, Matsuura Y and Hiraga K 1984 Permanent magnet materials based on the rare earth-iron-boron tetragonal compounds *Trans. IEEE* **MAG-20** 1584–1589

Schetky L M 1979 Shape memory alloys *Scientific American* 74–82

Smith D P E and Binnig G 1986 Ultrasmall scanning tunnelling microscope for use in a liquid helium storage dewar *Rev. Sci. Instrum.* **57** 2630–2631

Smith S T and Chetwynd D G 1990 An optimised magnet-coil force actuator and its application to linear spring mechanisms *Proc. Inst. Mech. Eng.* **204** 243–253

Smith S T and Chetwynd D G 1992 *Foundations of Ultraprecision Mechanism Design* (London: Gordon and Breach)

Squire P T 1990 Phenomenological model for magnetization, magnetostriction and ΔE effect in field-annealed amorphous ribbons *J. Magn. Magn. Mater.* **87** 299–310

Tai Y and Muller R S 1989 IC-processed electrostatic micromotors *Sensors and Actuators* **20** 49–55

Tojo T and Sugihara K 1985 Piezoelectric driven turntable with high positioning accuracy *Bull. Japan. Soc. Prec. Eng.* **19** 135–137

Trimmer W and Gabriel K 1987 Design considerations for a practical electrostatic micromotor *Sensors and Actuators* **11** 189–206

Tritt T M, Gillespie D J, Kamm G N and Ehrlich A C 1987 Response of piezoelectric bimorphs as a function of temperature *Rev. Sci. Instrum.* **58** 780–783

Uchino K, Cross L E and Nomura S 1980 Inverse hysteresis of field induced elastic deformation in the solid solution 90 mol% $Pb(Mg_{1/3}Nb_{2/3})O_3$–10 mol% $PbTiO_3$ *J. Mater. Sci.* **15** 2643–2646

Uchino K, Nomura S, Cross L E, Newham R E and Jang S J 1981 Review: Electrostrictive effect in perovskites and its transducer applications *J. Mater. Sci.* **16** 569–578

van de Leemput L E C, Rongen P H H, Timmerman B H and van Kempen H 1991 Calibration and characterisation of piezoelectric elements as used in scanning tunnelling microscopy *Rev. Sci. Instrum.* **64** 989–992

van Randeraat J and Setterington R E 1974 *Piezoelectric Ceramics* (London: Mullard)

Yaeger J R 1984 A practical shape-memory electromechanical actuator *Mechanical Engineering* 51–55

Young R, Ward J and Scire F 1972 The Topographiner: An instrument for measuring surface microtopography *Rev. Sci. Instrum.* **43**(7) 999–1011

Yuan Z, Feng Z and Zhao W 1987 Quick responding magnetostrictive microfeeder *Proc. 6th Int. Conf. on Precision Eng. Osaka* 593–598

7

Silicon Microactuators

Mehran Mehregany

Miniaturization of mechanical systems promises unique opportunities for progress of science and technology in new directions. Micromechanical devices and systems are inherently smaller, lighter, faster, and usually more precise than their microscopic counterparts. However, development of micromechanical systems requires appropriate fabrication technologies which enable:

definition of small geometries;
precision dimensional control;
design flexibility;
interfacing with control electronics;
repeatability, reliability, and high yield;
low cost per device.

Clearly, successful miniaturization of electronics would not have been possible without the integrated circuit (IC) fabrication technology, which in fact meets all of the above criteria. For example, one of the primary reasons for the widespread application of IC technology is the low cost per device.

It is therefore natural that IC fabrication technology (or microfabrication) has so far been the primary enabling technology for the development of micromechanical systems. Microfabrication provides a powerful tool for batch processing and miniaturization of mechanical systems into a dimensional domain not accessible by conventional (machining) techniques. Furthermore, microfabrication provides an opportunity for integration of mechanical systems with electronics to develop high-performance closed-loop-controlled microelectromechanical systems (MEMS). In the most general form, MEMS would consist of mechanical microstructures, microsensors, microactuators, and electronics integrated in the same environment.

Integrated fabrication techniques, which are possible with IC fabrication technology, eliminate the need for discrete component assembly, which is not practical for the fabrication of MEMS. Hence, further dimensional control, including component size and intercomponent clearance, is limited only by the processing technology.

Even though miniaturization of mechanical systems is directly compared with that for electronics, two important points should be noted. First, it is not implied that all mechanical systems will benefit from miniaturization. In fact, for many mechanical systems (such as an automobile) miniaturization is not practical. However, microfabricated sensors and actuators which enable performance improvement will be increasingly integrated into such conventional macroscopic mechanical systems. Miniaturization and application of microtransducers for monitoring is justified when the performance-to-cost ratio is improved. Second, the IC-based fabrication technologies are inherently planar, not allowing full flexibility for three-dimensional (3D) design. A mature technology for micromechanical systems will require complementary fabrication techniques which provide 3D design capabilities.

With a finite effort and a limited space, one cannot hope to achieve a comprehensive summary or a thorough survey of the entire field of MEMS. Instead, this chapter provides a summary overview of the silicon microactuator technology. A discussion of the origin and development of the field is presented, followed by a review of silicon micromachining technology and materials aspects. Microactuation mechanisms and representative microactuators are discussed. Examples are selected from the available literature as well as from research activities of the author to demonstrate the important points. Micromotors, which are submillimetre-sized electric actuators capable of unrestrained motion in at least one degree of freedom, are selected to provide a microactuator case study for a more in-depth discussion of microactuator concepts.

7.1 ORIGIN AND DEVELOPMENT OF THE FIELD

One of the initial applications of IC technology to the fabrication of micromechanical structures originated from efforts to develop solid-state pressure transducers in the mid-1960s. These pressure microsensors used diffused piezoresistors to obtain an electronic signal proportional to the strain produced by the deformation of a microfabricated thin silicon diaphragm subjected to pressure. Over the next two decades, the field matured substantially, and a variety of microsensors for measuring position, velocity, acceleration, pressure, force, torque, flow, magnetic field, temperature, gas composition, humidity, pH, solution/body fluid ionic concentration, and biological gas/liquid/molecular concentrations were developed and some commercialized.

Prior to 1987, silicon micromachining had been used to fabricate a variety of micromechanical structures such as thin silicon diaphragm beams, and other suspended structures [1] in single-crystal silicon or in films deposited on a silicon substrate. These micromechanical structures were generally limited in motion to small deformations and were physically attached to the substrate. Such elastic components could be used as flexible joints, but their overall usefulness in the design of *mechanisms* was limited. 'Mechanism' as used here is a means for transmitting, controlling, or constraining relative movement and is considered as a collection of rigid bodies connected together by joints. During 1987 and 1988, a turning point in the field was reached when, for the first time, techniques for integrated fabrication of mechanisms on silicon were demonstrated [2, 3]. It was now possible to fabricate mechanical parts that could execute unrestrained motion in at least one degree of freedom (e.g. gears, gear trains, linkages, etc). Shortly thereafter, this technology enabled the development of electrostatic micromotors [4–7] and motivated the development of other types of microactuators such as valves, pumps, switches, tweezers, and lateral resonant devices.

The recent progress in microactuators is transforming the conventional field of solid-state transducers into what is increasingly becoming known as MEMS. Rapid advances in device design, fabrication, materials, testing, packaging, and applications, as well as the rapid expansion of the field has brought about many symposia and conferences to report on the progress being made [8]. The field of solid-state transducers has traditionally been application driven and technology limited, and has emerged as an interdisciplinary field which involves many areas of science and engineering. Microactuators, and MEMS in general, have followed a similar trend.

7.2 SILICON MICROMACHINING

Microfabrication technology, or more specifically silicon micromachining, has been a key factor for the rapid development of microactuators. Silicon micromachining has emerged as an extension of IC technology. It is a combination of silicon IC technology and precise etching, bonding, deposition, interconnections, and packaging. Micromachining is used to fabricate a variety of precise 3D microstructures of great diversity including diaphragms, bridges, cantilever beams, needles, gears, springs, grooves, orifices, pyramids, complex mechanical suspensions, linkages and micromotors. These microstructures combined with electronics have been used successfully to realize a wide range of solid-state sensors and actuators.

Bulk and surface micromachining, as well as bonding of substrates and electroforming in conjunction with x-ray lithography, play an important role in the fabrication of MEMS. Bulk micromachining is the technique used to shape the bulk of the silicon substrate from all sides to fabricate desired 3D microstructures. On the other hand, surface micromachining, developed more recently, is the technique of fabricating microstructures by processing on the surface of a silicon wafer. This is achieved by successive deposition and etching of structural and sacrificial layers on the surface of the silicon wafer to construct the desired micromechanical device. Bonding is the technique of attaching a silicon wafer to another substrate (e.g. silicon, glass, etc.), allowing added freedom in the 3D design of microstructures. Finally, x-ray lithography can be used to define very high aspect ratio features in a photosensitive material, which is then used as a mould for electroforming metallic microstructures. This technique is now known as the LIGA process.

7.2.1 Bulk Micromachining

Bulk micromachining of silicon uses wet and dry etching techniques in conjunction with etch masks and etch stops to sculpt micromechanical devices from the silicon substrate. There are two key capabilities that make bulk micromachining of silicon a viable technology. First, anisotropic etchants of silicon such as potassium hydroxide (KOH), ethylene-diamine and pyrocatecol (EDP), and hydrazine are available which preferentially etch single-crystal silicon along different crystal planes. Second, etch masks and etch-stop techniques are available which can be used in conjunction with silicon anisotropic etchants to selectively prevent regions of silicon from being etched. As a result, it is possible to fabricate microstructures in a silicon substrate by appropriately combining the etch mask and etch-stop patterns with the anisotropic etchant capabilities.

Extensive reviews of bulk micromachining by anisotropic etching of silicon have been published [1, 9, 10]. Figure 1 demonstrates the basic concepts of bulk micromachining by anisotropic etching of a (100) silicon substrate. Anisotropic etchants of silicon etch the (100) and (110) planes significantly faster than the (111) planes, and etch-rate ratios of over 100/1 for (100) and (110) to (111) planes are possible. For example, for a (100) silicon substrate, etching proceeds along the (100) planes, while it is practically stopped along the (111) planes. Since the (111) planes make a 54.7° angle with the (100) planes, the slanted walls shown in figure 1 result. Due to the slanted (111) planes, the size of the etch mask opening determines the final etch result (e.g. a hole or a cavity). If the etch mask openings are rectangular (or square) and the sides are aligned with the [110] direction (i.e. the direction of the intersection line between (100) and (111) planes), no undercutting of the etch mask feature takes place, assuming the etch rate

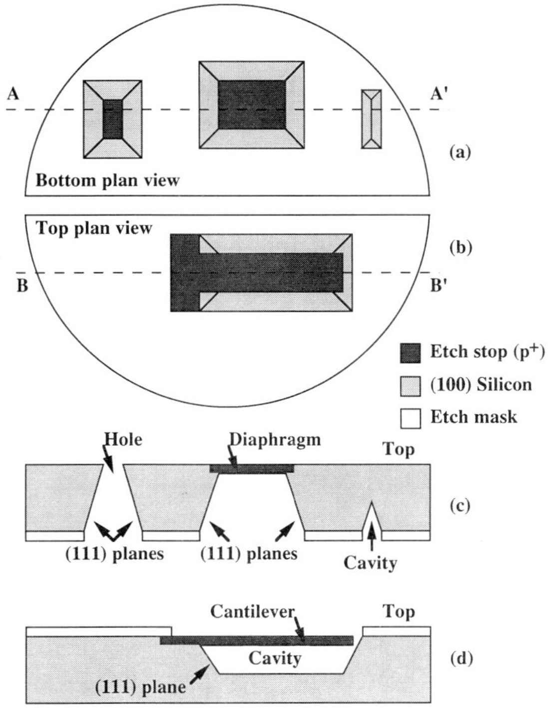

Figure 1. A schematic drawing demonstrating the basic concepts of bulk micromachining for a (100) silicon wafer: (a) bottom plan view of a wafer; (b) top plan view of a wafer; (c) cross-section A–A′; and (d) cross-section B–B′. (a) and (c) demonstrate etching from the back through openings defined in the etch mask. (b) and (d) demonstrate underetching of convex corners from the front side.

of the (111) planes to be negligible. For convex corners, misalignment with the [110] direction, and curved edges in the etch mask openings, significant underetching may occur until the etch is limited by the (111) planes. As shown in figure 1, the underetching characteristics can also be utilized to easily fabricate free-standing microstructures. Figure 2 shows a bulk micromachined silicon cantilever which is fabricated by undercutting of the cantilever convex corners of the beam geometry (which is defined by an etch stop) from the front side of a wafer (see figure 1). The slanted (111) planes can also be seen in the photograph.

Silicon dioxide, silicon nitride, and some metallic thin films (e.g. chromium, gold, etc) provide good etch masks for typical silicon anisotropic etchants. As shown in figure 1, these films are used to effectively mask areas of silicon that are to be protected from etching and to define the

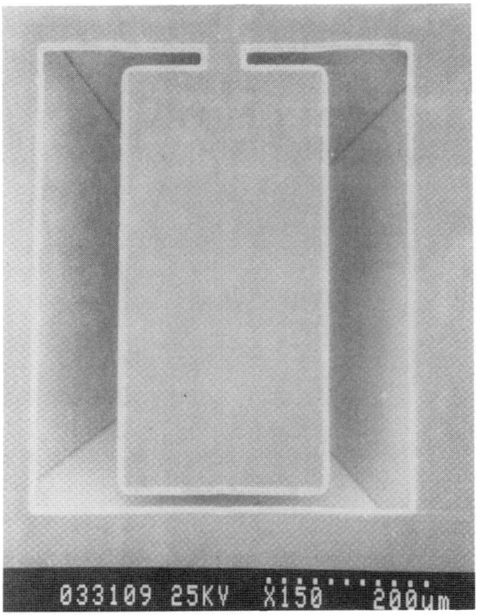

Figure 2. A 5 μm thick p^+ silicon cantilever beam fabricated by bulk micromachining described in figure 1(b), (d).

initial geometry of the regions to be etched. In terms of etch stops, two techniques have been widely used in conjunction with silicon anisotropic etching. Heavily boron-doped silicon (above 7×10^{19} cm^{-3}), referred to as p^+ etch-stop, is effective in practically stopping the etch. Alternatively, the pn-junction technique can be used to stop the etch when one side of a reverse-biased junction diode is etched away. As shown in figure 1, silicon etch stops often form the micromechanical device that is eventually delineated by the etch.

In addition to wet-chemical anisotropic etching of silicon, bulk micromachining may also use wet-chemical isotropic etchants such as solutions based on hydrofluoric (HF) and nitric acids [1]. However, the utility of these etchants for precision micromachining is limited since the etch contours are often controlled by the etching conditions rather than by the crystalline structure of the silicon substrate.

Dry etching has advanced significantly in the past decade because of its critical role in fine-line pattern delineation in the IC technology. For the same reason, dry etching has now become an integral part of micromachining technology. However, dry etching requirements for micromachining are typically much different from those in IC technology [1]; in many cases, they are much more demanding because of the larger depth of etches involved. Deep etches (e.g. up to several hundred microns) require the development of new etch recipes with acceptable etch rates, appropriate etch masks, techniques for improved profile control, reliable end-point detection, and provisions against excessive heating of the surface. Some studies have addressed the use of deep etching (i.e. 40–50 μm) of single-crystal silicon for bulk micromachining applications [11, 12]. Figure 3 shows an SEM photograph of a turbine rotor etched into a silicon wafer to a depth of 45 μm using an SF_6–CCl_2F_2 etch chemistry.

7.2.2 Surface Micromachining

Surface micromachining, demonstrated by Nathanson *et al* [13] in building a free-standing metal-gate field-effect transistor, relies on encasing specific device structural parts in layers of a sacrificial material during the fabrication process. The sacrificial material is then dissolved away in a chemical etchant that does not attack the structural parts. The basic concepts for surface micromachining are demonstrated schematically in figure 4. Surface micromachining requires a compatible set of structural materials, sacrificial materials, and chemical etchants [14]. The structural materials need to possess the desired electrical and mechanical properties for the application in mind. The sacrificial materials need to possess the proper mechanical properties so as not to cause device failure during the fabrication process. The chemical etchants must be able to preferentially etch the sacrificial

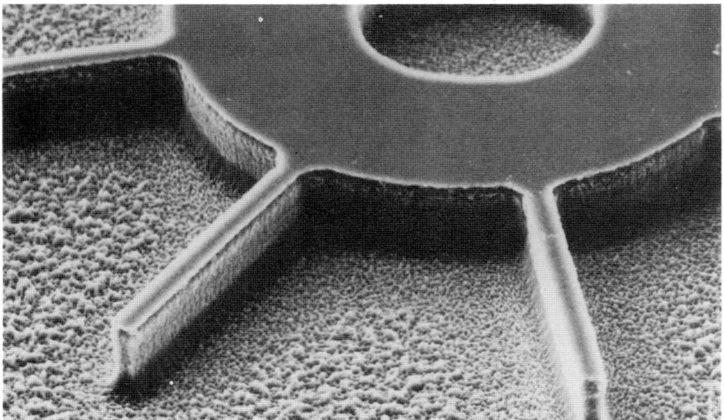

Figure 3. An SEM photograph of a 600 μm diameter single-crystal silicon turbine rotor fabricated by dry etching [12].

materials with respect to the structural parts, must have proper viscosity and surface tension characteristics to enable them to fully remove the sacrificial layers, and must not leave residues behind.

For monolithic micromechanical systems, it is desirable to have structural materials with good mechanical properties including no residual stress, very high modulus, high fracture strain, very low frictional coefficient, and high wear resistance. Due to the actuator requirements, good conductors and good insulators are required for the structural materials; however, it is more desirable to be able to tailor the material conductivity in the range of these two extremes. Finally, the material candidates should be compatible with IC technology.

Polysilicon surface micromachining using doped or undoped polysilicon as the structural material and silicon dioxide as the sacrificial material has been the most widely used surface micromachining technique. The potential of the polysilicon/oxide material system for micromachining was first demonstrated by Howe and Muller in the development of a resonant microbridge vapour sensor [15, 16] using a process similar to that in figure 4. Using polysilicon surface micromachining, complex micromechanical devices such as gear-trains, turbines, cranks, tweezers and more complex mechanisms have now been fabricated on silicon. Furthermore, polysilicon surface micromachining has been the backbone of the fabrication technology for electrostatic microactuators. To exemplify the potential of

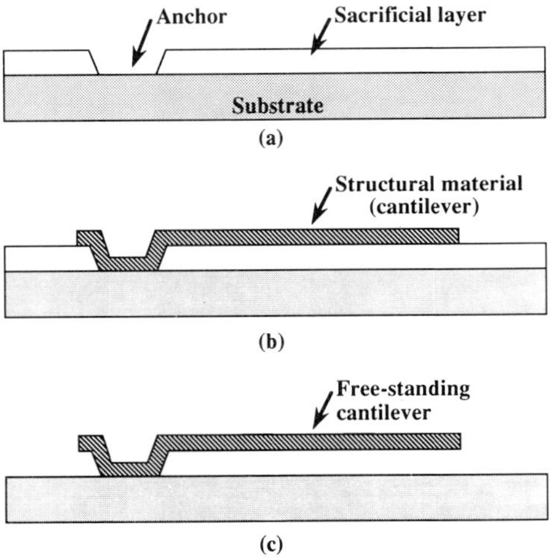

Figure 4. A schematic drawing demonstrating the basic concepts of surface micromachining: (a) after deposition and patterning of the sacrificial layer; (b) after deposition and patterning of the structural layer; and (c) releasing the cantilever by etching away the sacrificial layer.

surface micromachining, a process for fabricating a rotor on a centre-pin bearing is described next.

The basic centre-pin bearing fabrication process uses two polysilicon and two silicon dioxide depositions as well as four photolithography steps [2, 3]. Figure 5 shows cross-sectional views of a rotor on a centre-pin bearing during the fabrication process. After a blanket deposition of silicon dioxide, the bushing moulds are time etched to a desired depth in the oxide (figure 5(*a*)). After a blanket deposition of polysilicon, the rotor is patterned (figure 5(*b*)) by reactive-ion etching (RIE). Note that the bushings are formed automatically. The bushings are an integral part of the centre-pin bearing design and eliminate rotor-to-substrate stiction problems in the final device [2]. After a second blanket deposition of silicon dioxide, the bearing anchor is opened (figure 5(*c*)). After a second blanket deposition of polysilicon, the centre-pin bearing is defined (figure 5(*d*)). At this point, the sacrificial silicon dioxide is dissolved in HF to release the rotor. When released, the bushing rests on the silicon substrate, reducing the surface area of contact between the rotor and the substrate. During operation, the bushing slides on the substrate. Figure 6 is an SEM photograph of a gear-train

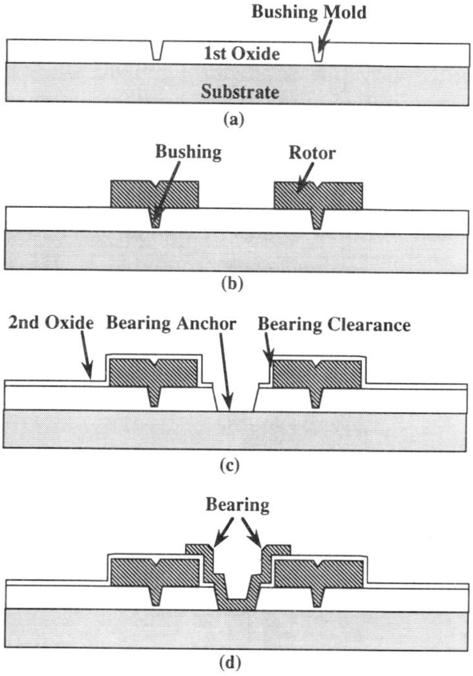

Figure 5. The general centre-pin fabrication process: (a) after the bushing mould is patterned; (b) after the rotor is patterned; (c) after the bearing anchor is patterned; and (d) completed device.

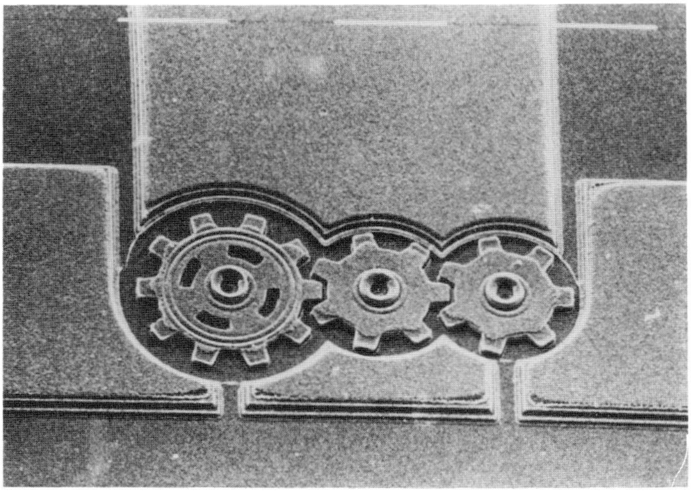

Figure 6. A polysilicon gear train fabricated with the centre-pin process. The smaller gears are 125 μm in diameter, and the larger gear is 175 μm in diameter [2].

fabricated with the centre-pin process. The first and second polysilicon layers are 4.5 μm and 3 μm thick, respectively. The first and second sacrificial oxide films are 4 μm and 1 μm thick, respectively. The bushing mould is etched to a depth of 2 μm.

The centre-pin process has the capability of producing a self-aligned or a non-self-aligned bearing. Let s_1 and s_2 denote the thicknesses of the first and second oxide sacrificial layers, respectively. Let p_1 and b denote the thickness of the first polysilicon layer and the height of the bushing. Then if $p_1 + b < s_1 + s_2$, the rotor can slide under the bearing, resulting in a non-self-aligned bearing. The bearing clearance for the non-self-aligned case is the difference between the radius of the top of the bearing anchor opening and the rotor inner radius. This clearance can be large and is governed by the mask dimensions as well as the slope of the bearing anchor sidewalls produced by the fabrication process. If $p_1 + b > s_1 + s_2$, the bearing clearance is specified by the oxide thickness on the sidewalls of the rotor inner radius. This results in a self-aligned bearing, which is the type used for all of the centre-pin bearing micromotors to date.

A variation of the centre-pin bearing process for fabricating a flange bearing has been described [3] and is shown in figure 7. The overall flange bearing process is very similar to the centre-pin process and uses only minor modifications to accomplish the flange bearing. For the same number of film depositions, the flange process provides more flexibility in design compared with the centre-pin process. However, the centre-pin process is somewhat

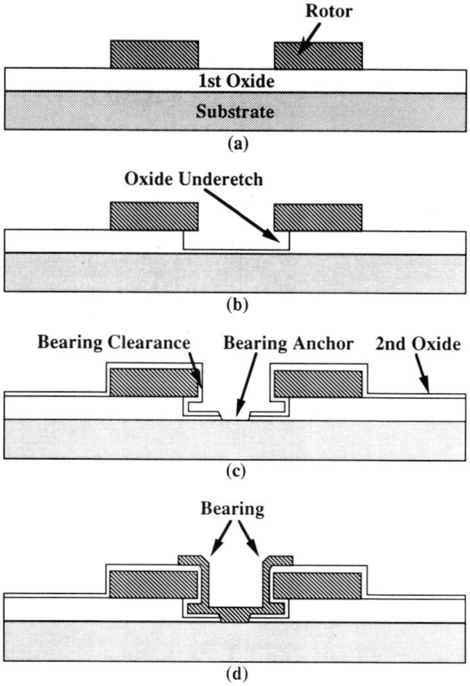

Figure 7. The general flange fabrication process: (a) after the rotor is patterned; (b) after the flange mould is patterned; (c) after the bearing anchor is patterned; and (d) completed device.

simpler. Figure 8 shows an entirely released three-degrees-of-freedom mechanism fabricated with the flange process. The first and second polysilicon layers are 2.2 μm and 1 μm thick, respectively. The first and second sacrificial oxide layers are 2.3 μm and 0.5 μm thick, respectively. The flange extends by 1.5 μm under the first polysilicon layer components.

Dry-etching techniques for precision pattern delineation in 1 μm to 5 μm thick films play an important role in surface micromachining. For many surface micromachined electrostatic actuators, the etch profiles obtained by dry etching can significantly affect final device performance. For example, in the micromotors described later in this chapter, excessive roughness of the dry etched rotor inner radius and air gap sidewalls leads to locking of the bearing during operation and premature electric field breakdown in the gap upon large electric excitations, respectively.

7.2.3 Bonding

Bonding techniques provide a means for attaching a silicon substrate to another substrate, which is usually silicon or glass, to provide added device

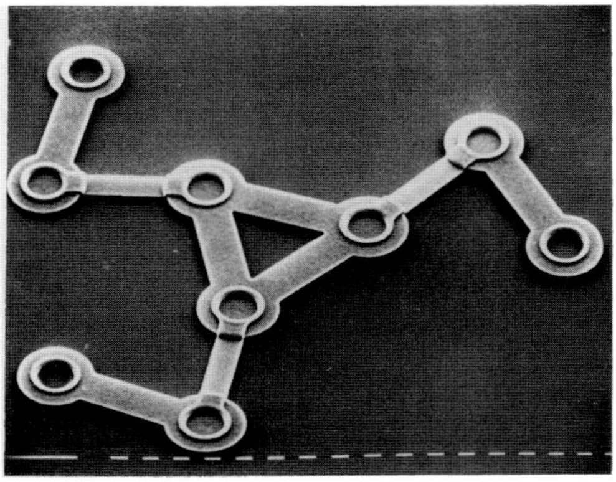

Figure 8. A released three-degrees-of-freedom planar polysilicon mechanism [17]. Each link is 100 μm long. The three outer revolute joints are attached to the substrate while the six inner revolute joints are free to slide on the substrate. The triangular output platform can move in the plane of the substrate. Each white dash is 10 μm wide.

design flexibility, mechanical support, electrical connection, and thermal sink or isolation. Techniques for electrostatic bonding of silicon to glass substrates have been widely used in microfabrication of sensors [18]. Bonding of a silicon to a glass substrate in conjunction with bulk micromachining has also been used to fabricate electrostatic microactuators [19]. An important recent development has been silicon fusion bonding (SFB) which has become the fabrication technology base for current commercially available pressure sensors for industrial and medical applications [20, 21]. SFB is the technique of bonding two silicon wafers together at a high temperature (i.e. near 1000 °C). It is simple and does not require the application of an external pressure, electric field, or an intermediate bonding layer. One indication of the integrity of the bond obtained by SFB has been the excellent performance of pn-junctions formed by bonding a p-type wafer to an n-type one as a large current rectifier.

SFB provides a new means for constructing 3D structures with excellent thermal stability and mechanical strength. For example, using silicon fusion bonding, ultra-stable, high-temperature piezoresistive pressure sensors have been fabricated which were not possible based on conventional techniques [22]. Additionally, the added flexibility in 3D design provided by SFB enables the fabrication of new and more complex devices. Figure 9 shows an example of a microvalve fabricated by fusion bonding of two silicon wafers. Figure 10 shows a microfabricated pump made by bonding four silicon wafers. Operational characteristics of these devices are described later in this chapter.

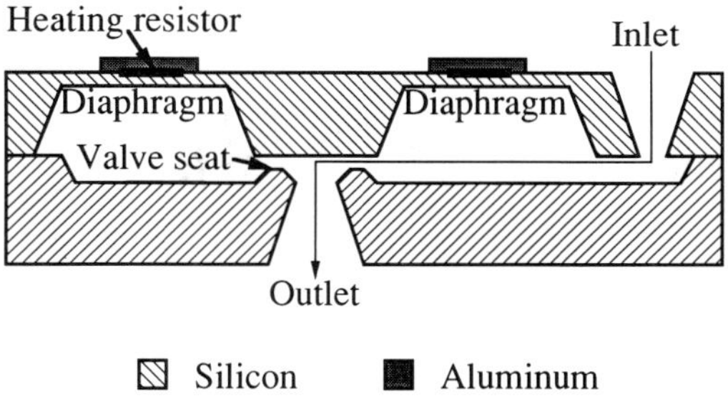

Figure 9. Schematic cross-sectional drawing of a bimetallic diaphragm microvalve fabricated by silicon fusion bonding of two wafers [23].

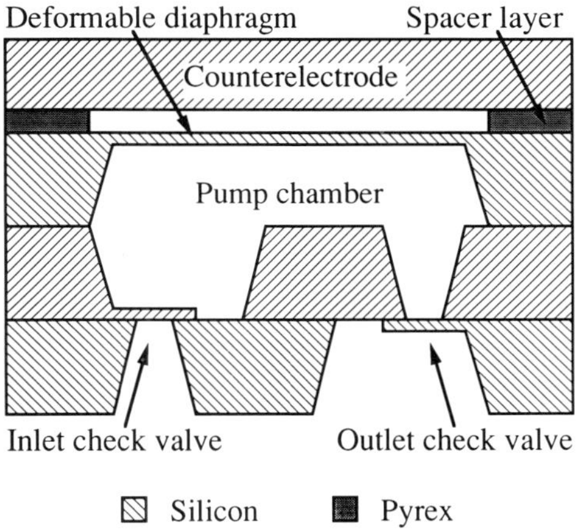

Figure 10. Schematic cross-sectional drawing of an electrostatic diaphragm micropump fabricated by bonding of four silicon wafers [24].

7.2.4 The LIGA Process

An inherent characteristic of the IC fabrication technology is the planar geometry. In general, fabrication of high aspect ratio features which have a height at least an order of magnitude larger than width is not readily possible. The planar nature of microfabrication technology limits the design of mechanical devices, including many silicon microactuators. In the last five years, significant progress has been reported in the development of process technologies for high aspect ratio microstructures. New x-ray photolithography techniques have been developed which allow definition of features with very large (e.g. nearly 100) height-to-width aspect ratios. Once these features are formed in the photosensitive material, plating techniques are used to plate up metallic (e.g. nickel) microstructures using the resist as a mould. This technique was first pioneered in Germany and has become known as the LIGA process [25, 26]. In the LIGA process, a synchrotron radiation source is used for exposure; features as small as a few microns wide and as deep as several hundred microns with vertical sidewalls have been defined. Other variations based on the LIGA process are now being developed [26] in hopes of approximating LIGA-like features without using a synchrotron radiation source.

7.3 MATERIALS ASPECTS

Availability of an extensive materials base which can meet the requirements for MEMS development is essential for future progress of the field. Additionally, this materials base must be accompanied with a well-documented materials data base which enables accurate device design. The materials development needs of the field are well recognized but are at preliminary stages [27]. In the past decade, there have been a number of advances in exploring new materials and studying material properties for microsensors and microactuators. Much of this research is documented in the journals and conference proceedings of the field [8]. In addition to single-crystal silicon, polysilicon, silicon dioxide, and silicon nitride, other materials are being explored for MEMS. Ceramic and polymeric materials represent important opportunities for new or improved devices, especially in ultrasonic and optical areas. Shape memory alloy metals (such as TiNi), CVD tungsten, CVD diamond-like films, polymeric piezoelectrics, polyimides, and high-temperature superconductive materials are other interesting examples. Figure 11 shows a microgear fabricated from sputtered amorphous SiC.

As important as developing new materials is a thorough understanding of the mechanical properties of materials that are being used. The mechanical properties of interest include residual stress, Young's modulus, Poisson's ratio, yield stress, ultimate stress and fatigue. Table 1 compares the mechanical properties of silicon and silicon derivatives with other more conventional mechanical materials. Many of the data in table 1 are from bulk measurements; the mechanical properties of the material in thin-film form can be somewhat different [27]. The goal of mechanical properties studies in MEMS (and in general in thin films) is to develop models which relate the process parameters to the film microstructure and mechanical properties. In the past decade, a number of techniques have been developed for or have been applied to the measurement of mechanical properties of films used in the fabrication of microsensors, microactuators, and MEMS

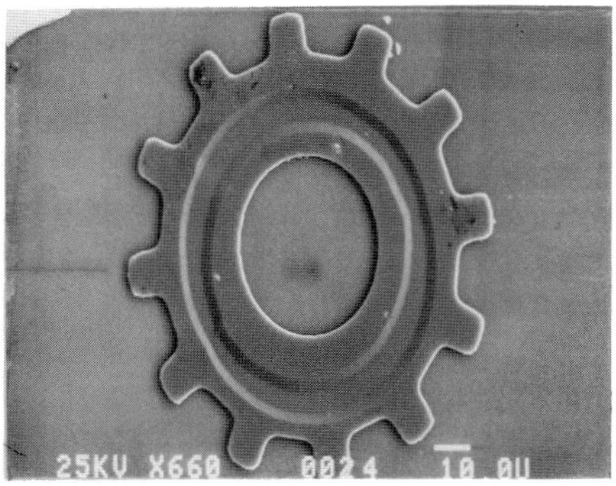

Figure 11. A 100 μm diameter microgear fabricated from amorphous thin-film SiC deposited on an oxidized silicon wafer [28].

Table 1. A comparison of mechanical properties of silicon and silicon derivatives with more conventional mechanical materials. (Reproduced from [1].)

	Yield strength (10^{10} dyne cm^{-2})	Knoop hardness (kg mm^{-2})	Young's modulus (10^{12} dyne cm^{-2})	Density (g cm^{-3})
[a]Diamond	53	7000	10.35	3.5
[a]SiC	21	2480	7.0	3.2
[a]TiC	20	2470	4.97	4.9
[a]Al_2O_3	15.4	2100	5.3	4.0
[a]Si_3N_4	14	3486	3.85	3.1
[a]Iron	12.6	400	1.96	7.8
SiO_2 (fibres)	8.4	820	0.73	2.5
[a]Si	7.0	850	1.9	2.3
Steel (max strength)	4.2	1500	2.1	7.9
W	4.0	485	4.1	19.3
Stainless steel	2.1	660	2.0	7.9
Mo	2.1	275	3.43	10.3
Al	0.17	130	0.70	2.7

[a] Single crystal.

[27, 29, 30]. These techniques have been used for measurement of mechanical properties in silicon, polysilicon, silicon dioxide, silicon nitride, polyimides and metals. Figure 12, for example, shows a micromachined structure used to measure the fracture strain of a polyimide thin film. Since the residual stress in the polyimide thin film is tensile, contraction of the wider portion of the microstructure upon release pulls on the necks and can produce fracture if the geometry is selected appropriately.

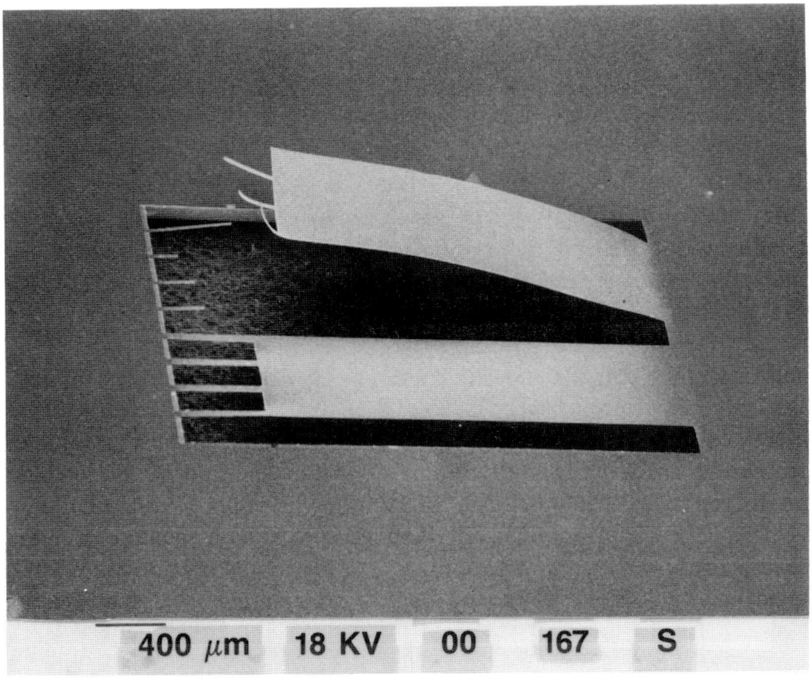

Figure 12. A bulk micromachined polyimide microstructure for measuring the fracture strain of the film [30]. The microstructure with the wider necks has remained intact while the other has fractured upon release from the substrate. By making an array of microstructures of varying geometry the fracture strain can be determined.

Even with silicon, there remain mechanical properties to understand. For example, heavily boron-doped silicon to above 7×10^{19} cm^{-3}, which is known as p^+ silicon, is used extensively as an etch stop in conjunction with anisotropic etchants of silicon to fabricate mechanical elements. However, p^+ silicon suffers from residual stresses that can affect the performance of a device fabricated by p^+ etch-stop techniques. For p^+ silicon, the measurement of residual stress and modulus have been made using various microfabricated devices [31–36]. The results of such measurements show that p^+ silicon is typically under a residual tensile stress (near 60 MPa), which is related to the boron doping as well as the dislocations and other defects which occur in the lattice. The p^+ silicon modulus, 108 ± 20 GPa, in the (110) direction is about 40% smaller than that for undoped single-crystal silicon. On the other hand, buckling of p^+ diaphragms has also been observed, which would indicate a net residual compressive stress in the film. The fatigue of p^+ cantilever beams, under repeated bending with generated stresses well below the yield stress, has been observed by the shift of the resonant frequency [37]. A clear understanding of the observed mechanical properties of p^+ silicon is not available.

In addition to the mechanical properties of the materials used in microsensors and microactuators, the interface characteristics, namely friction and wear properties, are becoming increasingly important. Friction and wear characteristics of the thin-film materials under consideration can have a profound impact on the potentials and limitations of microelectromechanical devices such as micromotors and micromechanisms. Many of the friction and wear effects in micromechanical devices are expected to be different from their macroscopic counterparts because of the larger surface-to-volume ratio, higher smoothness of the contact surfaces, and significantly smaller forces. A number of techniques have been developed to study friction and wear in microelectromechanical devices [14]; a few of these techniques are discussed later in this chapter in conjunction with micromotors.

7.4 MICROACTUATION MECHANISMS

A critical need for the progress of microactuator technology is the development of actuation forms which are compatible with the materials and processing technologies of silicon microelectronics. Additionally, actuation is to be powered and controlled electrically, allowing full utilization of integration with on-chip electronics. Despite these stringent requirements, numerous physical phenomena have recently been demonstrated for microactuator applications. Commonly used microactuation methods have so far been either electromagnetic or thermal [38, 39]. Most notable among demonstrated electromagnetic microactuators are electrostatic, piezoelectric, and magnetic; among thermal microactuators are bimetallic, thermo-pneumatic, and shape memory alloy.

As will be seen later this chapter, microactuators can be divided into two mechanical classes: (i) mechanisms and (ii) deformable microstructures. Mechanism-type microactuators (e.g. micromotors) provide displacement/force through rigid-body motion while deformable microstructures (e.g. beams and diaphragms) provide displacement/force through mechanical deformation (or straining). While electromagnetic microactuation has been demonstrated in conjunction with both classes of microactuators, thermal microactuation is primarily suitable for deformable microstructures. Figure 13 shows schematic examples of deformable thermal microactuators. These devices can be fabricated by bulk micromachining of silicon using a p^+ etch stop as described earlier in this chapter. Heating of the bimetallic cantilever through power dissipation in the heating resistor causes the metal and silicon layers to expand, resulting in deflection of the cantilever tip due to the

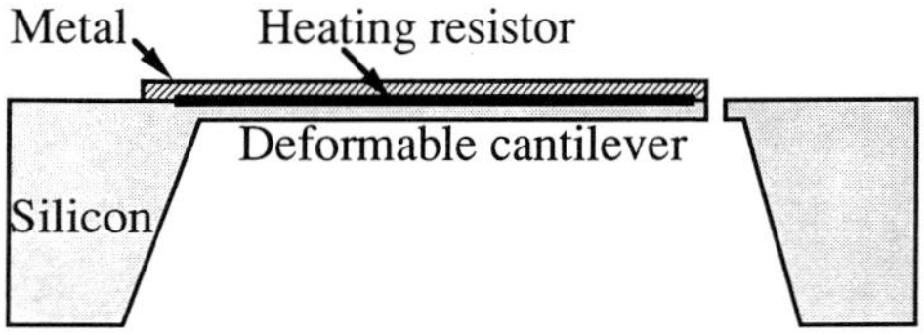

(a)

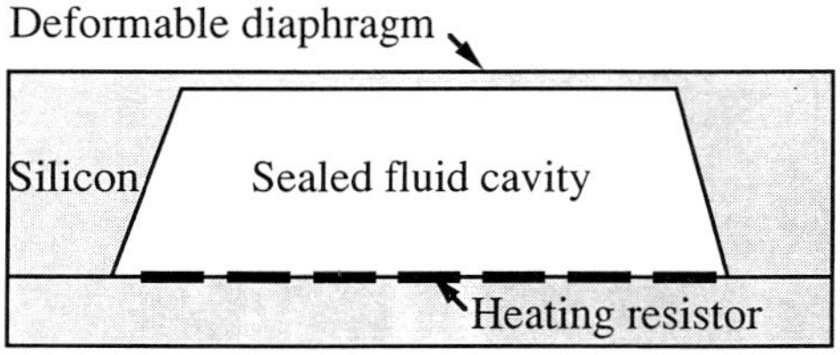

(b)

Figure 13. Schematic cross-sectional drawing of basic deformable thermal microactuators: (a) a bimetallic cantilever geometry; and (b) a thermopneumatic diaphragm geometry.

mismatch in the thermal expansion coefficient of the two layers. In the case of the thermopneumatic devices, heating of the fluid in the cavity will result in expansion of the fluid volume and deformation of the diaphragm.

Shape memory alloys (SMAs) are metals which have shape recovery characteristics. When these alloys are plastically deformed at one temperature, they will completely recover their original shape on being raised to a higher temperature [40]. In recovering their shape, the alloys can produce a displacement or force, or a combination of the two, as a function of temperature. The development of thin-film SMAs provides the potential for development of silicon-based MEMS which incorporate SMA microactuators. The specific SMA that has been studied for microactuator applications is Nitinol or TiNi, an alloy of titanium and nickel [41, 42]. The reasons for selecting TiNi over other SMAs for MEMS applications have been its (i) superior SMA characteristics, (ii) better compatibility for IC processing, (iii) high resistivity allowing resistive heating, (iv) ability to provide transition temperatures over the wide range of −273 °C to 100 °C by varying the nickel–titanium ratio and adding other elements, and (v) ability of TiNi to provide two-way memory.

Figure 14 shows schematic examples of cantilever and diaphragm deformable microstructures but with piezoelectric and electrostatic actuation. Note that these devices can also be fabricated by bulk micromachining

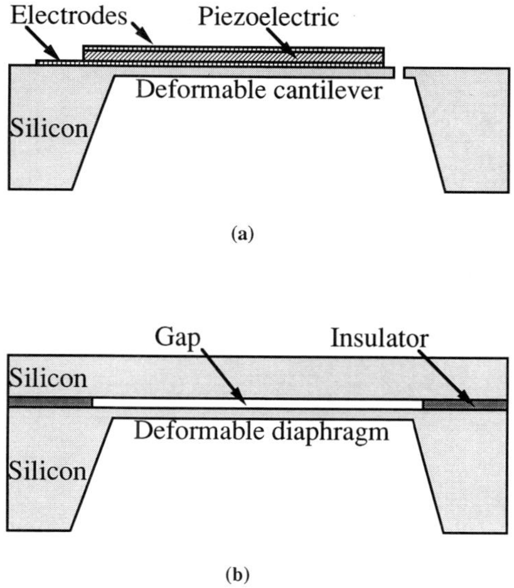

Figure 14. Schematic cross-sectional drawing of basic deformable electric microactuators: (a) a piezoelectric cantilever geometry; and (b) an electrostatic diaphragm geometry.

of silicon using a p^+ etch stop. For the piezoelectric device, application of a voltage across the thickness of the piezoelectric layer will cause the piezo to contract or expand depending on the polarity of the applied voltage. The change of length of the piezo layer with applied voltage causes the cantilever to deflect. For the electrostatic device, application of a voltage across the gap attracts the diaphragm toward the counter electrode. These simple devices exemplify the operation of deformable electric and thermal microactuators.

In addition to deformable microstructures, MEMS require large-motion actuators (e.g. micromotors) to power the linkages. Electric, as opposed to magnetic, drive has been argued as the preferable actuator drive in the microscopic scale [43]. This is mainly for two reasons. First, the electric breakdown limit in air increases drastically for microscopic (order of 1 μm) air-gap sizes. Typically, the electric field breakdown limit in air is near 3×10^6 V m^{-1} for macroscopic dimensions and increases to above 10^8 V m^{-1} for microscopic air-gap sizes [14, 43, 44]. However, for magnetic actuators, the magnetic saturation limit is independent of actuator size. Therefore, for microscopic sizes, the stored energy density limits (discussed in the next paragraph) for electric and magnetic actuators become comparable and are of the order of 10^5–10^6 J m^{-3}. Second, required materials for electric actuators, namely conductors and insulators, are available in the IC industry while those for magnetic actuators (i.e. magnetic materials) seldom are. However, more recently a magnetic micromotor made of nickel has been demonstrated by using the LIGA fabrication process [45]. The fabrication process for this device is not fully integrated which puts it at a disadvantage in comparison with the electrostatic polysilicon micromotors described in the next section.

It is often of interest to compare the various microactuation methods for their relative advantages and disadvantages. However, detailed comparisons are only accurate when performed in the light of an application. Nevertheless, it is possible to outline some general points when comparing various microactuation methods. An important point of interest for a microactuation method is the amount of force (or mechanical energy) that can be generated. One approach for general comparison is to estimate the amount of energy, W, available per unit volume. For example, the energy per unit volume, W, available from a TiNi SMA material can be calculated from the area under the operating curve on the stress–strain diagram. For an austenitic state with a yield strength of 420 MPa and an available shape memory strain of 8%, W is 4×10^7 J m^{-3}. For bimetallic microactuators, W would depend on the magnitude of temperature change but can be of comparable magnitude. For an electrostatic microactuator $W = \frac{1}{2}\varepsilon_o E_{max}^2$ is nearly 4×10^5 J m^{-3}, assuming an electric field breakdown limit (E_{max}) of 3×10^8 V m^{-1} for microscopic air gaps. (ε_o is the permittivity of free space.) For piezoelectric microactuators, W may be up to 10^5 J m^{-3}, assuming lead zirconate titanate (PZT) to be the piezoelectric material. For magnetic microactuators, $W = \frac{1}{2}(B_{max}^2/\mu_o)$ is about 10^6 J m^{-3}, assuming a magnetic flux density saturation of 1.5 T. (μ_o is the permeability of free space.)

Other factors must also be considered in comparing different microactuation methods. In general, thermal microactuators have a slow response time (e.g. on the order of tens of milliseconds) and have high power consumption (e.g. on the order of tens of milliwatts). Comparatively, electromagnetic microactuators can be much faster (e.g. microseconds response time) and consume far less power, particularly electric microactuators. The construction of thermal microactuators often requires the final free-standing part to be a laminate of layers with very different mechanical properties. This often complicates device design as evidenced by the fact that reported bimetallic actuators [46, 47] exhibit a preset deflection due to the residual stresses in the thin films. Piezoelectric microactuators require deposition of additional film since silicon is not piezoelectric. Furthermore, static operation of piezoelectric microactuators is limited by charge leakage. Electrostatic microactuators can be fabricated by using conducting and insulating films which are common to microelectronics technology. Static excitation of electrostatic microactuators requires voltages across insulating gaps and nearly no loss. Magnetic microactuators require magnetic materials which are not common in IC technology and often require some type of manual assembly. Static excitation of magnetic microactuators requires current through windings and persistent conduction losses.

7.5 MICROMOTORS: A CASE STUDY IN MICROACTUATOR TECHNOLOGY

This section concentrates on the electric micromotor technology. The micromotors under discussion are surface micromachined using heavily phosphorus-doped polysilicon by low-pressure chemical vapour deposition (LPCVD) for the structural material, deposited oxide for the sacrificial layers, and LPCVD silicon nitride for electrical isolation. In general, in an electric actuator, the attractive and repulsive forces generated by electric charge distributions are used to convert electrical to mechanical energy. By proper commutation of these charge distributions on a set of stationary electrodes, known as the stator, and a set of moving electrodes, known as the rotor, continuous motion of the rotor can be achieved. Similarly to their magnetic counterparts, electric motors can be categorized based on the specific mechanism of electric actuation. For example, electric induction motors are the counterparts of magnetic induction motors while permanent electret motors are the counterparts of permanent magnet motors.

To date, only rotary variable-capacitance micromotors (which are the counterpart of variable reluctance magnetic motors) have been reported because of their design and material requirement simplicity. The operation of these micromotors relies on the storage of electrical energy in a variable rotor-stator capacitance. The change in this capacitance in the direction of motion is proportional to the output torque of the micromotor. In terms of design, one aims for optimum torque generation by enhancing the capacitance change in the direction of motion. In terms of material requirements, only good conductors and insulators are required. The material requirements are compatible with polysilicon surface micromachining which has been used almost exclusively for micromotor fabrication thus far. However, other fabrication processes are now starting to emerge [48, 49]. Rotary instead of linear designs have been emphasized thus far because rotary motors provide force-to-torque leverage. Furthermore, rotary motion is more suitable for motor diagnostic and characterization studies. For example, rotary micromotors can be spun with an air jet as well as electricity.

7.5.1 Micromotor Designs

Reported micromotors have thus far been based on variable-capacitance designs. For the sake of brevity, we will drop the variable-capacitance term in describing the micromotors below. Two micromotor designs based on a top-drive and a side-drive configuration have been reported [14]. Initially, the top-drive design was considered because of its potentially superior motive torque characteristics. However, top-drive micromotors demonstrated significant design and fabrication shortcomings, namely, rotor instability problems and planarization requirements [6]. These shortcomings motivated the development of side-drive micromotors [4–7]. The side-drive design eliminates the rotor instability problems and the need for planarization. However, this is achieved at the expense of sacrificing motive torque.

Figure 15 is an SEM photograph of a typical top-drive polysilicon micromotor. This micromotor is fabricated using a two-level centre-pin polysilicon process identical to that in section 7.2.2 with one added step. The added step is a blanket deposition of a LPCVD silicon nitride isolation layer on the starting silicon substrate. The silicon nitride addition is required for electrical isolation of the stator poles. Note that the stator is fabricated from the second polysilicon deposition and does not require additional process steps. For the micromotor in figure 15, both polysilicon layers are 2.5 μm and are heavily doped with phosphorus. The sacrificial layers are deposited from low-temperature oxide (LTO). The first LTO thickness is 2.3 μm, while the second LTO thickness is 1 μm; the bushing height is 1.8 μm.

In the top-drive micromotor design, electric energy is stored in the air gap which is typically 1–2 μm and is formed by the overlap of the stator and rotor poles. The air gap in the top-drive micromotor is defined by the two LTO thicknesses and the rotor bushing depth, all of which are easy parameters to control. The micromotor operation relies on tangential electrostatic forces tending to align the rotor poles under the excited stator

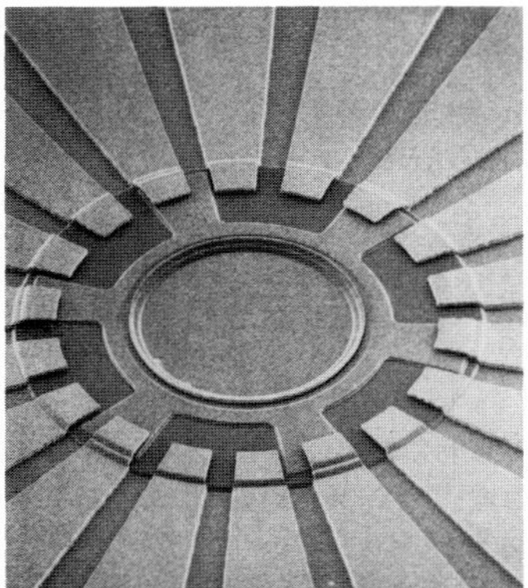

Figure 15. An SEM photograph of a 300 μm diameter, salient-pole, centre-pin, top-drive micromotor.

poles. For micromotor operation, the substrate is electrically grounded while a three-phase bipolar excitation is applied to the stator. In bipolar excitation, opposite stator poles of the same phase are excited with equal voltage magnitudes but opposite polarities.

Due to its significantly larger rotor-stator capacitance changes with rotor position, the top-drive micromotor design can potentially provide a larger motive torque compared with the side-drive designs presented below. However, associated with the rotor-stator and the rotor-substrate capacitances, there are large vertical forces tending to clamp the rotor up to the stator poles or down to the substrate. In the former case, the micromotor will be destroyed by the large current passing through the rotor due to the shorting out of the stator poles. In the latter case, the micromotor will not operate due to the large frictional forces associated with rotor-substrate clamping. Rotor instability is a severe limitation of this design.

Furthermore, since the stator poles are not all located over rotor poles during device fabrication, they are not at the same height (see figure 15). To fix this problem, an excessively thick second LTO deposition and a subsequent planarization would be required. Otherwise, the stator poles fabricated over rotor poles are at a higher elevation than the remaining stator poles. The planarization requirement is another severe limitation of the top-drive micromotor design. These difficulties have proved the top-drive micromotor design to be impractical at the present time. Therefore, research has focused on the design, fabrication, and characterization of salient-pole and harmonic (also known as wobble) side-drive micromotors. These side-drive micromotor designs overcome the design and fabrication difficulties associated with the top-drive micromotors, but at the cost of sacrificing output torque.

The salient-pole and harmonic side-drive micromotors are hereafter referred to as salient-pole and wobble micromotors. For the sake of brevity, the side-drive term is dropped since all *operational* micromotors to date have been of the side-drive design. We will, however, use the term side-drive micromotors to refer to both the salient-pole and wobble micromotors collectively. Figure 16 is an SEM photograph of a three-phase, salient-pole, polysilicon micromotor, and figure 17 is an SEM photograph of a wobble polysilicon micromotor. These micromotors are fabricated using the basic centre-pin process of section 7.2.2 with some additional steps. Their fabrication process is described in the next section. It is also possible to fabricate these micromotors with flange bearing designs [50].

In the side-drive micromotors, electric energy is stored in the gap formed between the rotor and stator sidewalls. Note that capacitance changes with rotor position are significantly smaller for the side-drive micromotors compared with the top-drive design above. This results in a significantly smaller motive torque for the side-drive micromotors. In contrast to the top-drive design, in which the air-gap dimension is controlled by deposited film

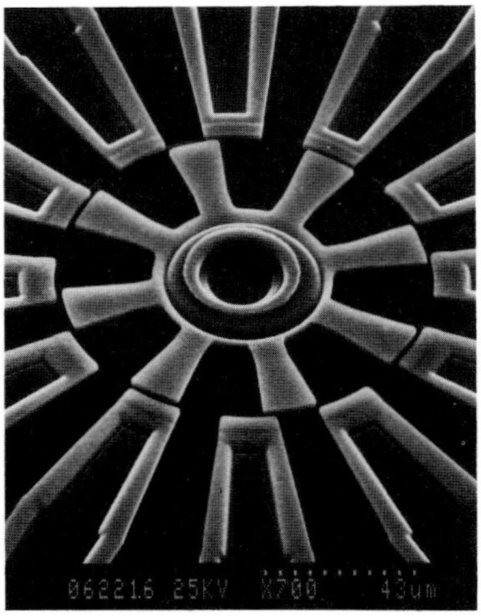

Figure 16. An SEM photograph of a 100 μm diameter, salient-pole, centre-pin, side-drive micromotor; the rotor/stator gap is 1.5 μm wide.

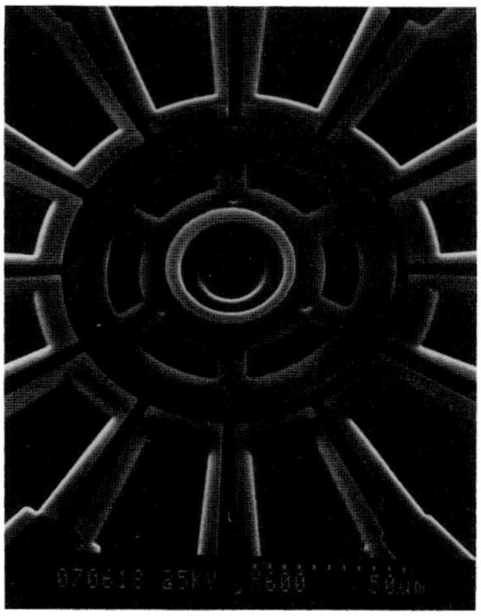

Figure 17. An SEM photograph of a 100 μm diameter, harmonic, centre-pin, side-drive micromotor; the rotor/stator gap is 2.5 μm wide.

thicknesses, the air-gap dimension in the side-drive micromotors is controlled by photolithography and pattern delineation; hence, it is more difficult to control.

A conductive polysilicon shield beneath the rotor and in electrical contact with it has been suggested [51, 52] to be effective in electrostatically shielding the rotor from the substrate as well as providing a means for biasing the rotor. Figure 18 is a schematic cross-sectional view of the micromotors in figures 16 and 17, demonstrating the polysilicon shield incorporated beneath the rotor. During motor operation, the bushing slides over the shield. This mechanical contact is intended to provide the required electric contact between the rotor and the shield.

For the salient-pole micromotor operation, the shield is electrically grounded while a three-phase bipolar excitation is applied to the stator. Micromotor operation relies on tangential electrostatic forces tending to align the rotor poles with the excited stator poles. The salient-pole micromotors have been fabricated [52, 53] with stator-to-rotor pole number ratios of 3:1, 3:2 (similar to that in figure 16), and 2:1. The former two require three-phase excitation while the latter requires two-phase excitation. The 3:2 architecture provides superior torque coverage in comparison with the other two architectures [52, 53].

Operating principles of wobble motors have been described in detail [50, 52, 54, 55]. The central feature of the wobble micromotor is that the rotor wobbles around the centre bearing post. A plan view schematic of the wobble micromotor design is shown in figure 19. For micromotor operation, the rotor is electrically grounded by externally grounding the shield, and the stator poles are excited sequentially. Micromotor operation relies on normal

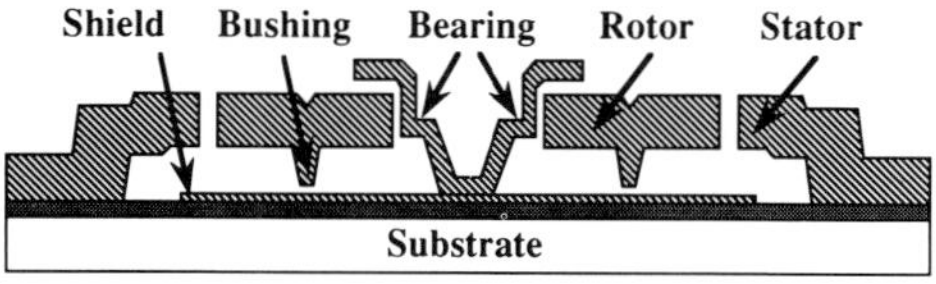

Figure 18. Schematic cross-sectional view of the micromotors in figures 16 and 17.

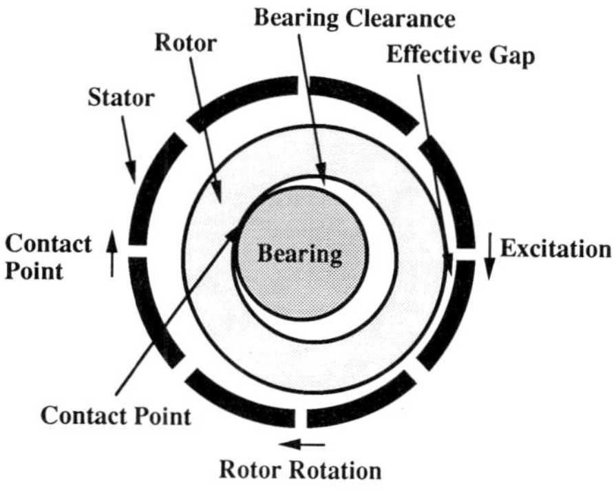

Figure 19. Plan view schematic of the wobble micromotor. For simplicity, this plan view schematic only shows eight stator poles.

electric forces attracting the rotor to the excited stator pole. The centre bearing limits lateral rotor motion and prevents it from contacting the excited stator pole and shorting out. By ensuring a bearing clearance smaller than the nominal air-gap size, the need for insulation between the rotor and the stator is eliminated. The rotor lateral movement (i.e. wobble distance) is equal to the clearance in the bearing which is the difference between the bearing and inner rotor radii. As the electric excitation travels around the stator poles, the rotor rolls on the bearing. Since the rotor circumference is larger than that of the bearing, the rolling action also produces rotor rotation. The rotor rotational speed is proportional to the electric signal frequency by the gear ratio. The gear ratio also relates the angle of rotation of the stator excitation to the angle of rotation of the rotor. In this design, the rotor lateral movement (i.e. wobble distance) is equal to the clearance in the bearing. For the wobble micromotor in figure 17, the bearing clearance is 0.3 μm while the air gap is 2.5 μm.

For comparable geometrical designs, the wobble micromotors provide higher motive torque than the salient-pole micromotors. This results from the fact that the operation of the wobble micromotors relies on normal electric forces of attraction rather than the tangential component in the salient-pole micromotors. Additionally, the inherent gear ratio in the wobble micromotors is a significant advantage. However, in the wobble micromotors, the rotor wobbles as it turns which may be a disadvantage in some applications.

7.5.2 Fabrication Issues

Micromotor fabrication has almost exclusively been based on the centre-pin or flange processes. Figure 20 is a schematic description of a centre-pin bearing micromotor fabrication process based on the author's work. This fabrication process requires three polysilicon depositions, two silicon dioxide depositions, and one silicon nitride deposition along with six photolithography steps. The micromotors of figures 16 and 17 were fabricated with this process. Detailed descriptions of micromotor fabrication have been documented [52, 53, 56].

Initially, substrate isolation consisting of a sandwich of 1 μm silicon-rich silicon nitride over 1 μm thermal oxide is established (figure 20(a)). Next, a 0.35 μm thick polysilicon layer is deposited, heavily doped with phosphorus, and patterned to form the shield (figure 20(a)). The first sacrificial LTO layer, 2.3 μm thick, is deposited by CVD and the bushing mould openings are defined. The bushing moulds are time etched in buffered oxide etch (BOE) to a depth of 1.8 μm (figure 20(a)). Note that the remaining 0.5 μm LTO in the bushing mould trenches result in a corresponding rotor-stator vertical offset in the final device. The LTO layer is then patterned and etched to open the stator anchors, exposing the nitride layer below (figure 20(a)).

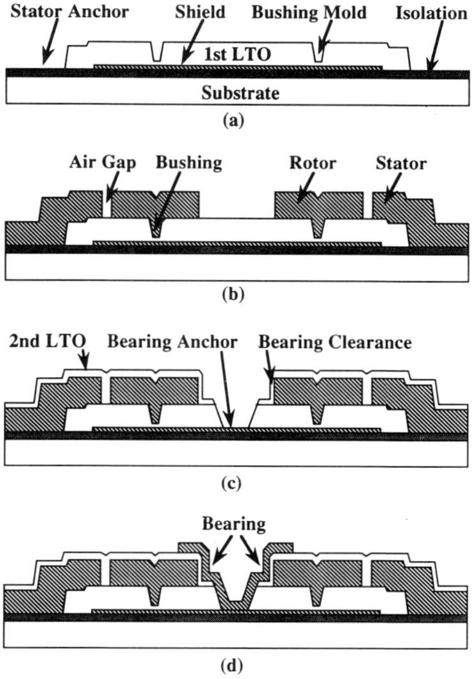

Figure 20. Centre-pin bearing micromotor fabrication process: (a) after the bushing mould and stator anchor are patterned; (b) after the rotor and stator are patterned; (c) after the bearing anchor is patterned; and (d) completed device.

A 2.5 μm thick polysilicon layer is deposited, heavily doped with phosphorus, and patterned to form the rotor and stator (figure 20(*b*)). The rotor-stator pattern definition and delineation step is the most critical in the micromotor fabrication process since it defines the rotor/stator gap size. We have been able to routinely fabricate 1.5 μm air gaps in the 2.2 μm thick rotor-stator polysilicon of our side-drive micromotors. Next, a second sacrificial LTO layer is deposited which provides 0.3 μm of LTO coverage on the rotor and stator sidewalls and nearly 0.5 μm of LTO coverage on the top surfaces (figure 20(*c*)). The 0.3 μm LTO coverage on the sidewalls corresponds to the bearing clearance in the micromotor. Next, the bearing anchor is defined and etched in the LTO sacrificial layers, exposing the electric shield below (figure 20(*c*)).

A 1 μm thick polysilicon layer is deposited, heavily doped with phosphorus and patterned to form the bearing (figure 20(*d*)). At this point, the completed device is immersed in a 1:1 HF:deionized water solution to dissolve the sacrificial LTO and release the rotor (figure 18). Proper micromotor release is not only required for micromotor operation, but also affects micromotor performance [52, 56]. However, our current knowledge of the effect of the release process on the micromotor operational characteristics is limited. The effect of release and operational environment on micromotor performance requires further research to identify the competing mechanisms that affect friction, stiction, and failure in these devices. To reduce the final HF release time, a sufficient number of holes is patterned in the rotor to expedite the HF underetch by reducing the underetch distance.

7.5.3 Operational Characteristics

The overall operational characteristics of the side-drive micromotors reported to date are similar. However, the specific performance characteristics vary for the different micromotors. Both centre-pin and flange bearing salient-pole and wobble micromotors have been demonstrated [50]. Micromotor operation has been demonstrated in dielectric liquids (e.g. deionized water and silicone oil) [57], as well as in gaseous environments (e.g. nitrogen, argon, oxygen, and room air) [50]. In general, the operational excitation voltages range from a few tens of volts for some micromotors [52] to just over 100 V for others [4, 51, 58, 59]. For the 1 μm to 3 μm rotor/stator gaps used in the micromotors, the above excitation voltages correspond to electric field intensities above 10^8 V m^{-1}, which corroborates the initial estimates of breakdown in microscopic air gaps as discussed in section 7.4. The motive torque of the side-drive micromotors is typically of the order of tens of piconewton metres (pN m).

Micromotor operation has only been studied under open-loop excitation. Under open-loop operation, a square-wave electric excitation is commutated on the stator; the stator phases are turned on one at a time. The excitation switching time is chosen to be larger than the settling time of the rotor transient. Therefore, micromotor operation has been of a stepping rather than continuous motion. For the different salient-pole micromotors reported to date, both underdamped [60, 61] and overdamped [59] rotor transients have been observed for operation in gaseous environments. Our preliminary results suggest that the rotor transient is overdamped for operation in liquids due to the high viscous drag. The rotor transient settling time has typically been under 250 μs for salient-pole micromotors operating in a gaseous environment. For the salient-pole micromotors, open-loop stepping operational speeds up to 15 000 rpm have been reported for operation in a gaseous environment [52, 57]. For the wobble micromotors, the rotor transient motion has not been measured; however, the open-loop stepping operational speeds have been as high as 1000 rpm in room air [50]. Typical wobble micromotor gear ratios have been in the range of 70 to 200. The maximum attainable speeds for operation in gaseous environments have been limited by the power supply and not the micromotor.

For the micromotors reported to date, micromotor performance and lifetime has varied significantly. For the first operational salient-pole micromotor [4], only a very slow stepping operation could be achieved at excitation voltages above 100 V. This shortcoming was attributed to large electric forces attracting the rotor to the substrate. A conductive plate beneath the rotor and in electrical contact with it has been used in later designs to electrostatically shield the rotor from the substrate (as schematically demonstrated in figure 18). In these micromotors, both performance and lifetime have been improved significantly over time. Excitation voltages as low as 30 V have been sufficient in operating some of our current salient-pole and wobble micromotors [50, 57]. The open-loop operational speeds have been improved significantly as described above. Furthermore, micromotor lifetime has been extended to many millions of cycles over a period of several days for operation in room air or silicone oil [62, 57]. For these micromotors, a clear failure point has not yet been identified even though some micromotors are tested for a few hundred million start-stops. Finally, repeatable micromotor operation is often achieved in our current micromotors.

7.5.4 Friction

During micromotor operation, the dynamic mechanical contacts at the bearing and the bushing (or the flange) result in friction. Friction can present serious limitations to the performance and lifetime of a micromotor; it presents a load to the micromotor which directly opposes useful work. Micromotor failure may not necessarily be due to wear, but may often be due to increased friction beyond the micromotor drive torque. The bearing surface is primarily subject to rolling contact for a wobble micromotor and to sliding contact for a salient-pole micromotor. The bushing/flange surface is subject to sliding contact in both micromotor types. During wobble micromotor operation, the rotor is pulled against the bearing due to the normal electric force, and thus rolls on the bearing. In this case, the resulting friction at the bearing contact is critical to the micromotor operation. However, for the salient-pole micromotors, bearing frictional forces are detrimental to micromotor performance. For the salient-pole micromotors, when the rotor is electrically grounded and is symmetrically centred in between a set of excited stator poles, the resultant of the radial forces on the rotor due to each excited stator pole is zero. However, when the rotor displaces radially, a net radial electric force (commonly referred to as a side-pull force [59–61]) develops, which pushes the rotor into the bearing post. The side-pull force can result in significant friction in the bearing of salient-pole micromotors [59–61].

Several studies have contributed to the understanding of friction in micromotors and a review of the subject is available [14]. These studies have thus far focused almost exclusively on the friction of two polysilicon or polysilicon/silicon nitride surfaces sliding in contact with one another in gaseous environments. Differences between the frictional behaviour of centre-pin bearing salient-pole micromotors with and without silicon nitride coatings in their bearings have been reported. For example, the operational characteristics of the micromotors described in [50], which have two contacting polysilicon surfaces, are significantly different for operation in room air versus nitrogen. However, the micromotors described in [51, 58, 59], which have at least one silicon-nitride contacting surface do not demonstrate a significant dependence on the operational environment.

To study dynamic friction, different measurement approaches have been used. In one study [63], passive gears and turbines have been spun with an air jet which is then shut off abruptly to measure the slow down characteristics of the rotor. A simple dynamic model which accounts for viscous drag and coulomb friction is then used to estimate the dynamic coefficient of friction of the sliding contact surfaces. Dynamic coefficients of friction of 0.25 to 0.35 have been estimated for polysilicon/polysilicon sliding surfaces. In another study [60], a step excitation is used for actuating flange bearing salient-pole micromotors; the starting as well as stopping positions of the rotor are then measured. From many such measurements, the resulting data can be used in conjunction with a dynamic model to estimate the coefficient of friction for the sliding contact surfaces. In that study, coefficients of friction of 0.21 to 0.38 have been estimated for polysilicon/silicon nitride sliding surfaces. In another study, a stroboscopic dynamometry technique has been used to measure the details of rotor step transient in centre-pin salient-pole micromotors [52, 61]. The resulting data have been used in conjunction with a dynamic model to extract the value of frictional torque, which has typically been below 10% of the motive torque. Dynamic friction torque was also found to be independent of the bushing apparent area of contact [52,61]. The stroboscopic dynamometry technique has been used in another study [59] to estimate the dynamic coefficient of sliding friction for polysilicon on silicon nitride surfaces. The values reported are in agreement with those above from previous work [60].

The measurement of the dependence of the gear ratio on the applied electrical excitation in wobble micromotors has also been used to study friction and its dependence on the operational environment [50]. It has been found that the coefficient of friction for polysilicon on polysilicon is 0.38 in nitrogen and argon, 0.48 in oxygen, and 0.54 in room air (i.e. oxygen and moisture). These results are in agreement with other frictional studies using macroscopic samples [64]. It has also been found that the effect of operational environment on the performance of similar flange bearing micromotors is less significant since the flange is at a smaller radius than the bushing [50]. Furthermore, it is now well-documented that the bushing/

flange normal contact force is too large to be accounted for by the weight of the rotor alone [50, 61] and is partly of electrostatic origin.

To study static friction, studies have concentrated on estimating the motive torque necessary to initiate micromotor motion from rest [52, 65, 66]. This is accomplished by measuring the rotor position and the excitation voltage necessary to initiate motion from rest. These data are then used in conjunction with torque/position curves for the micromotor which are calculated from electric field modelling in the micromotor. Note that accurate estimates of the friction torque require an accurate micromotor torque/position analysis. In comparison with dynamic friction, static friction torques can be significantly larger [52, 65, 66]. For example, the resulting static friction torque resisting micromotor restart after 30 seconds at rest was found to be on the average near 50% to 70% of the motive torque for micromotors similar to those in which the dynamic friction torque was below 10% of the motive torque. The static friction torques were found to be independent of the bushing apparent area of contact in the experiments considered. Static friction torques exhibit a significant dependence on the duration of stationary contact. For micromotor restart after less than 1 second at rest, the frictional torque was a factor of two smaller than that for 30 seconds at rest [52, 65, 66]. Lateral resonant devices have also been used for static friction measurements [67]; the results are in agreement with the micromotor measurements.

7.5.5 Wear

Wear limits the lifetime of a micromotor in at least two ways. First, the bearing and bushings can wear down to the point where the micromotor ceases to function properly. This is particularly true of the salient-pole micromotors in which bearing wear leads to greater rotor side pull and hence greater friction and wear. Second, as wear particles are generated, they can accumulate in the bearing where they cause it to bind, or they can be drawn electrostatically into the rotor/stator gap where they accumulate and cause electric breakdown. Wear has been observed in passive polysilicon microfabricated gears and turbines using an air-jet drive [63]. In that study, polysilicon micromechanical components were typically spun at 300 000 rpm for an average of three minutes, and significant wear at the bearing and the bushing was observed in SEM inspections (see figures 21 and 22). However, quantitative wear measurements were not possible and the drive forces were not known. The first *in situ* quantitative measurements of wear in micromotors were reported by studying changes in the gear ratio of wobble micromotors under extended operation [62].

For the wobble micromotors, the effect of wear on micromotor operation can easily be measured *in situ*, providing a convenient means for studying wear under electric excitation. Mechanical wear in the bearing of the wobble

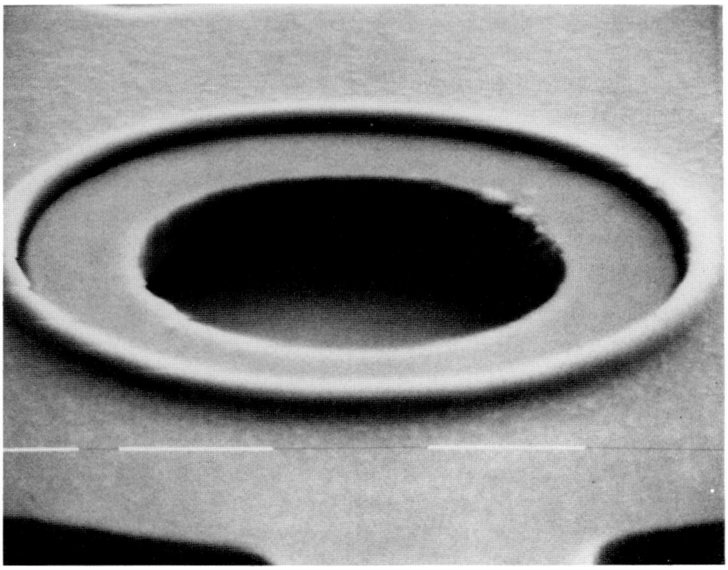

Figure 21. A SEM photograph of the rotor underside showing the circular bushing for a polysilicon rotor that is not operated [63]. Each white dash is 10 μm long.

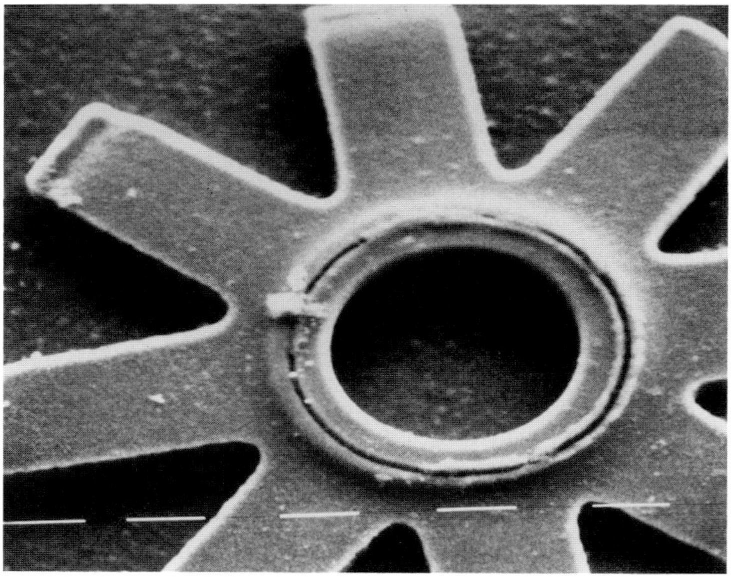

Figure 22. An SEM photograph of the rotor underside showing the wear of the bushing for a polysilicon rotor that is spun with an air jet at 2500 rps for five min [63]. Each white dash is 10 μm long.

micromotor results in a corresponding increase in the bearing clearance. Because the gear ratio of the wobble micromotor is equal to the bearing radius divided by the bearing clearance [62], under extended operation, changes in the gear ratio (reflected in changes in micromotor speed) can be a direct measure of bearing wear. This concept has been used to measure wear as a function of operational time (or number of cycles of rotation) in the bearing of the wobble micromotors. Extended operation of the wobble micromotors to near 100 million wobble cycles was studied [52, 62] at

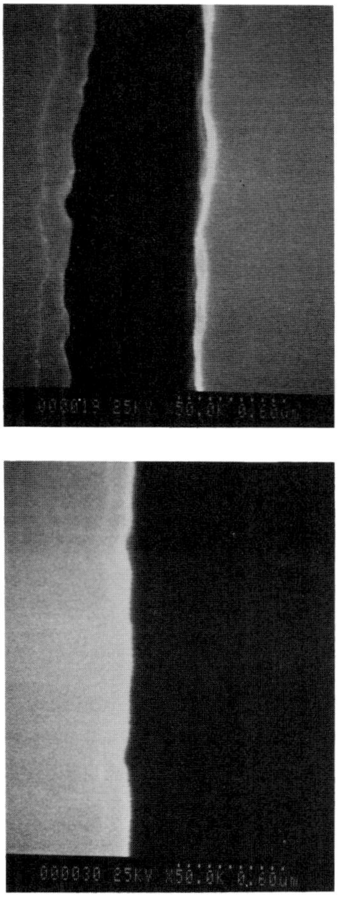

(a)

(b)

Figure 23. Magnified plan views from the rotor back side showing the rotor inner radius edges for two wobble micromotors [52, 62]: (a) an unoperated wobble micromotor where the rotor inner radius edge is on the left, the bearing edge is on the right, and the bearing clearance is the dark centre area; (b) a wobble micromotor after near 30 million wobble cycles where the rotor back surface is on the left.

excitation frequencies of 10 000 rpm and 25 000 rpm, for operational durations of 150 hours and 71 hours, respectively. The results indicate that bearing wear is important at the start of micromotor operation and results in changes in the gear ratio by as much as 20%. Typical gear ratios are near 90 at the start of wobble micromotor operation and decrease to near 70 as the bearing wears out. The wear in the bearing decreases substantially with time. This behaviour is attributed to the wearing out of the asperities of the rotor inner radius at the beginning stages of operation. These asperities result from the roughness of the RIE etched rotor inside radius sidewalls. Figure 23 shows the rotor inner radius from the backside for an unoperated wobble micromotor and one that has been operated for 30 million wobble cycles. The bearing wear is reflected in an increase of near 600 Å in the bearing clearance. The associated bearing contact forces in these experiments have been near 1 μN. Note that wear under electric excitation is substantially less than that for an air-jet drive described above. This is perhaps due to lower rotational speeds and smaller contact forces.

For the salient-pole micromotors, wear particles are observed to form in the bearing and at the bushings [52, 62]. When the salient-pole micromotors were operated at two rotor speeds at 160 and 2500 rpm for a total of 40 000 rotor revolutions, SEM inspections of the bushing surfaces and the rotor inner radius revealed scattered wear particles several hundred ångstroms in size. Wear was not detected on the polysilicon shield which the bushings slide over during micromotor operation. Micromotors which had not been operated, when dismantled for SEM inspection, exhibited clean surfaces with no particles. In general, based on the author's experience, wear is not presently a limiting factor in micromotor operation.

7.6 DEFORMABLE MICROSTRUCTURES

Another important class of microactuators relies on deformable microstructures to transmit force or torque. A few prominent microactuator types that exemplify the basic concepts are described here. Lateral resonant microstructures are the newest devices in the class of microactuators under discussion [68]; they are actuated by electrostatic forces. Figure 24 shows a schematic plan view of a typical lateral resonant microactuator. Polysilicon surface micromachining has thus far been the primary technique for the fabrication of these devices. Other fabrication techniques are emerging, however [19]. Lateral resonant structures typically consist of a moveable plate which is suspended a couple of microns above the substrate by a flexible folded beam support. The folded beam support is anchored (dark fill region in figure 24) to the substrate at one end and attached to the moveable plate at the other end. Electric forces through the use of comb-drive and/or

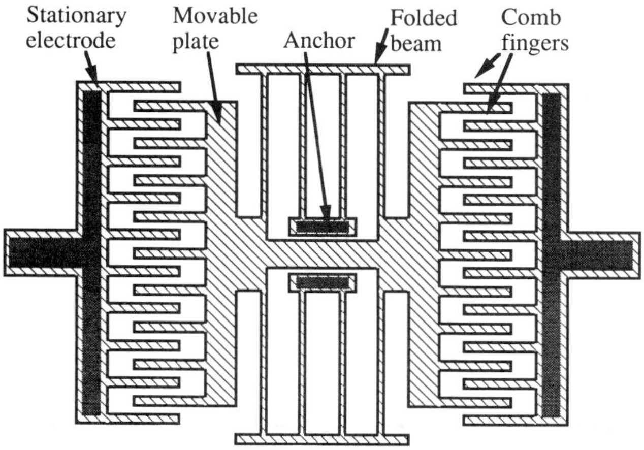

Figure 24. Plan view schematic of a typical polysilicon surface micromachined lateral resonant structure [68]. The supporting flexural beams are typically 80 to 180 μm long and have a 2 μm × 2 μm cross-section. The interdigitated electrode gap is typically 2 μm.

side-drive electrodes are employed to provide actuation. The suspension and actuation are designed for lateral deformation of the suspension and therefore lateral movement of the moveable plate. In resonant mode, these microactuators can provide lateral deformations near 10 μm; tens of kHz is typical of resonant frequencies for the designs reported [68]. In static mode, lateral deflections as high as 32 μm for 80 V excitations across 2 μm air gaps have been reported [69].

In addition to the suspension systems used in the lateral resonant structures, electric forces can be used in conjunction with more conventional microstructures such as beams and diaphragms (figures 10 and 14). For this group of microstructures, the deformation is predominantly in the direction normal to the substrate plane. For example, electric actuation in conjunction with a deformable diaphragm can be used to develop the micropump shown schematically in figure 10. This micropump employs two check valves which are simply cantilever beam flaps covering micromachined holes. When the counter electrode is excited, the diaphragm deflects up, increasing the pump chamber volume and reducing its pressure. The inlet check valve then opens as its cantilever flap bends up due to differential pressure. When the excitation is turned off, the diaphragm returns to its normal position, reducing the pump chamber volume and increasing its pressure. The outlet valve then opens allowing the fluid to exit. In the micropump described, the square diaphragm is 4×4 mm^2 and 25 μm thick; the actuator gap is 4 μm. Pumping has been demonstrated for actuation frequencies of 1 to 100 Hz. At 25 Hz, a pumping rate of 70 μl min^{-1} has been demonstrated when the outlet and inlet pressures are equal. Typical forward-to-reverse flow rate ratio of the check valves is 5000:1.

As described in section 7.4, another means of actuation is the use of the bimetallic effect [23, 46, 47]. Since microfabrication techniques lend themselves more easily to the stacking of the materials on top rather than beside each other, reported microactuators based on the bimetallic effect provide deformations that are predominantly in the direction normal to the plane of the substrate. For example, bimetallic cantilevers made of a sandwich of gold and silicon have been reported [46] which are similar to that shown schematically in figure 13. A polysilicon resistor is used to provide resistive heating of the gold/silicon sandwich in that study and deformations as high as 100 μm have been possible. Bimetallic actuation can be used in conjunction with a deformable diaphragm to develop the microvalve shown schematically in figure 9 [23]. In this case, heating of the bimetallic structure causes the diaphragm to deflect down closing the valve. The diaphragm in this device is 2.5mm in diameter and is 10 μm thick. A 5 μm thick aluminium layer is used as the metal component. Fully proportional control of flows in the range of 0 to 300 cc min^{-1} has been demonstrated with this valve for input pressures from zero to 100 psi. On/off flow ratios have been greater than 1000. To close the valve at 20 PSIG input, 1.5 W of power is required.

A number of other actuation mechanisms are also being studied. For example, the volume expansion of a gas upon heating has been used as a source of actuation [70] to develop thermopneumatic microactuators similar to that in figure 13. Piezoelectric films have been used to provide actuation in a variety of applications such as valves, pumps, positioning devices [71], and ultrasonic micromotors [72]. Typical piezoelectric thin films being used in the microactuator technology are zinc oxide (ZnO), PZT, and polyvinylidene fluoride (PVDF). Of these materials, PZT has the largest piezoelectric coefficients [73]. Examples of other less common microactuation techniques can be found in the journals and proceedings of the conferences of the field.

7.7 APPLICATIONS AND FUTURE TRENDS

The microactuator, and in general MEMS, technologies are in an infantile stage and are expected to mature over the next decade. A primary requirement for the maturation of the field is the development of applications with large commercial markets. Otherwise, the field will fail to fulfil its potential since industry support would be lacking. However, building on the foundation of the field which is based on microsensors and integrated circuits, the outlook is bright. The commercial markets for silicon pressure sensors, one of the first devices from the technology under discussion, are now well established. Most automobiles use at least one silicon pressure sensor (e.g. for measurement of pressure in the exhaust manifold). Other automotive applications of silicon pressure sensors are currently being pursued (e.g. measuring fuel tank, engine cylinder, and tyre pressure). On the medical side, disposable silicon pressure sensors are now being used for blood pressure monitoring. Even novelty applications such as pump-up athletic shoes are beginning to consider and use silicon pressure sensors. Such applications are being made possible due to the batch process nature of the technology which results in low unit cost.

Silicon accelerometers are finding their way into the automotive industry for active suspension and air bag release applications. For the air bag application, a self-testing feature is required to routinely test the accelerometer integrity. The self-testing feature requires integration of actuation with the sensor. In other words, the sensor must be active rather than passive. One version of such an accelerometer can be developed based on the lateral resonant device described in the previous section. The area of active sensors is expected to be a major application of the microactuator technology. In addition to being self-testing, active sensors can be operated in a force-balance mode to improve linearity and dynamic range.

Microvalves are envisioned to replace current solenoid valves in fuel injectors. Microvalves would require significantly less power to operate and would be less expensive. In fact, silicon micromachined nozzles are beginning to be used for fuel injector applications. Microvalves and micropumps can be used to develop fluid management microsystems for chemical analysis and long-term implantable drug delivery systems. The potential advantages are small size and low cost. However, biomedical applications are complicated due to biocompatibility and packaging requirements. The micromotors described above can provide an alternative technology to the larger bulk micromachined micropumps.

Other areas such as magnetic and optical information storage technologies are also envisioned to benefit from microactuator developments. Microactuators such as micromotors can be used for fine positioning of the magnetic or optical head, improving access time and storage density. However, it is likely that many of the more prominent applications of the technology under discussion will be forthcoming as the scientific and technical community is exposed to the MEMS technology. Furthermore, realization of many currently specified applications requires enabling technologies that are currently lacking. For example, there are needs for enabling technologies in micromachining technology, fabrication equipment, materials, high-force actuation, and modelling tools.

The current fabrication technology is inherently two-dimensional and restricts freedom in design of micromechanical devices. The fabrication equipment is optimized for IC technology in which the processing requirements are different from MEMS. For example, MEMS require thicker film deposition and etching, and often necessitate photolithography over high surface topography. The materials base in MEMS is primarily limited to silicon and its derivatives, limiting the advance of the technology. Current microactuation techniques either provide large motion or high force/torque. Large-motion, high-force/torque actuators are required for many practical applications. Finally, accurate modelling of sensors, actuators, processes, and material properties will be the basis for future research toward better devices and technologies. Despite its great importance, this topic was not addressed above due to a lack of space. However, efforts to develop computer-aided-design (CAD) tools and finite-element simulation programs for the MEMS area are underway [74, 75]. Most industrial designs of physical sensors are now based on detailed finite-element modelling of the mechanical microstructures using currently available software for conventional mechanics. Future progress in the areas outlined above is largely dependent on our ability to model the devices accurately.

The above topics can be exemplified in terms of future trends for micromotors. Even though the electric micromotor technology has progressed rapidly, a number of micromotor performance limitations still exist. The most important of these are small motive torque, friction problems, and mechanical coupling. The micromotor technology is expected to expand in several directions including new designs, materials, fabrication techniques, and applications. With the progress of the surface micromachining technology, it will become possible to implement new micromotor designs which improve motive torque and reduce friction. New materials may be used to improve the friction and wear characteristics of the micromotors. New micromotor designs must be developed that allow coupling of the rotor to other mechanisms; for example, incorporating micromotors in the three outer joints of the mechanism in figure 8 to develop an active x–y–θ table.

The importance of the systems approach to sensors and actuators will be increasingly recognized. The integration of signal conversion (i.e. transduction) and signal processing to achieve functional systems with better performance and lower cost or to create systems that have not been possible up to now will be explored in the near future. The integration of sensors, computing circuits, and actuators in a single fabrication process is a challenge that requires practical thinking in all phases of production of a useful system. For example, integration of micromotors, position sensors, and electronics with the mechanism of figure 8 could result in a precision positioning micromanipulator that could possibly find applications in optical component alignment.

REFERENCES

[1] Petersen K E 1982 Silicon as a mechanical material *Proc. IEEE* **70** 420–457

[2] Mehregany M, Gabriel K J and Trimmer W S N 1988 Integrated

fabrication of polysilicon mechanisms *IEEE Trans. Electron Devices* **ED-35** 719–723

[3] Fan L S, Tai Y C and Muller R S 1988 Integrated movable micro-mechanical structures for sensors and actuators *IEEE Trans. Electron Devices* **ED-35** 724–730

[4] Fan L S, Tai Y C and Muller R S 1988 IC-processed electrostatic micro-motors *Technical Digest, IEEE Int. Electron Devices Meeting (December 1988, San Francisco, CA)* 666–669

[5] Tai Y C, Fan L S and Muller R S 1989 IC-processed micro-motors: design, technology, and testing *Proc. IEEE Micro Electro Mechanical Systems Workshop (February 1989, Salt Lake City, UT)* 1–6

[6] Mehregany M, Bart S F, Tavrow L S, Lang J H, Senturia S D and Schlecht M F 1990 A study of three microfabricated variable-capacitance motors *Sensors and Actuators* **A21–23** 173–179

[7] Tavrow L S, Bart S F, Lang J H and Schlecht M F 1988 A LOCOS process for an electrostatic microfabricated motor *Sensors and Actuators* **A21–23** 893–898

[8] Journals: *IEEE Trans. on Electron Devices, J. Micromechan. Microeng., J. Microelectromechanical Systems, Sensor and Actuators,* and *Sensors and Materials*

International Conferences: *Int. Conf. on Solid-State Transducers (Biannual), IEEE Micro Electro Mechanical Systems Workshop (Annual).*

[9] Bassous E 1978 Fabrication of novel three-dimensional micro-structures by anisotropic etching of silicon *IEEE Trans. on Electron Devices* **ED-25** 1178–1185

[10] Bean K E 1978 Anisotropic etching of silicon *IEEE Trans. on Electron Devices* **ED-25** 1185–1193

[11] Fung C D and Linkowski J R 1985 Deep etching of silicon using plasma *Micromachining and Micropackaging of Transducers* eds C D Fung *et al* (Amsterdam: Elsevier)

[12] Mehregany M, Gabriel K J and Trimmer W S N 1987 Micro gears and turbines etched from silicon *Sensors and Actuators* **12** 341–348

[13] Nathanson H C, Newell W E, Wickstrom R A and Davis Jr J R 1967 The resonant gate transistor *IEEE Trans. Electron Devices* **ED-14** 117–133

[14] Mehregany M and Tai Y C 1992 Surface micromachined mechanisms and micromotors *J. Micromech. Microeng.* **1** 73–85

[15] Howe R T and Muller R S 1983 Polycrystalline silicon micro-mechanical beams *J. Electrochem. Soc.* **130** 1420–1423

[16] Howe R T 1984 *Integrated Silicon Micromechanical Vapor Sensor* PhD Thesis University of California, CA

[17] Behi F, Mehregany M and Gabriel K J 1990 A microfabricated three-degree-of-freedom parallel mechanism *Proc. IEEE Micro Electro*

Mechanical Systems Workshop (February 1990, Napa Valley, CA) 159–165

[18] Ko W H, Suminto J T and Yeh G J 1985 Bonding techniques for microsensors *Micromachining and Micropackaging of Transducers* ed C D Fung *et al* (Amsterdam: Elsevier)

[19] Gianchandani Y B and Najafi K 1992 A bulk silicon wafer dissolved process for microelectromechanical devices *J. Microelectromechanical Systems* **1** 77–85

[20] Backlund Y, Rosengren L, Hok B and Svedbergh B 1990 Passive silicon transducer intended for biomedical, remote pressure monitoring *Sensors and Actuators* **A21–A23** 58–61

[21] Petersen K, Barth P, Poydock J, Brown J, Mallon Jr J and Bryzek J 1988 Silicon fusion bonding for pressure sensors *Technical Digest, IEEE Solid-State Sensor and Actuator Workshop* (*June 1988, Hilton Head, SC*) 144–147

[22] Petersen K, Brown J, Vermeulen T, Barth P, Mallon Jr J and Bryzek J 1990 Ultrastable, high temperature pressure sensors using silicon fusion bonding *Sensors and Actuators* **A21–A23** 96–101

[23] Jerman H 1991 Electrically-activated, normally-closed diaphragm valves *Technical Digest, 6th Int. Conf. on Solid-State Sensors and Actuators* (June 1991, San Francisco, CA) 1045–1048

[24] Zengerle R, Richter A and Sandmaier H 1992 A micro membrane pump with electrostatic actuation *Proc. IEEE Micro Electro Mechanical Systems Workshop* (*February 1992, Travemunde, Germany*) 19–24

[25] Ehrfeld W, Bley P, Gotz F, Hagmann P, Maner A, Mohr J, Moser H O, Munchmeyer D, Schelb W, Schmidt D and Becker E B 1987 Fabrication of microstructures using the LIGA process *Proc. IEEE Microrobots and Teleoperators Workshop* (*November 1987, Hyannis, MA*)

[26] Guckel H, Christenson T R, Skrobis K J, Denton D D, Choi B, Lovell E G, Lee J W, Bajikar S S and Chapman T W 1990 Deep X-ray and UV lithographies for micromechanics *Technical Digest, IEEE Solid-State Sensor and Actuator Workshop* (*June 1990, Hilton Head, SC*) 118–122

[27] Senturia S D 1987 Can we design microbotic devices without knowing the mechanical properties of materials? *Proc. IEEE Microrobots and Teleoperators Workshop* (*November 1987, Hyannis, MA*)

[28] Tong L, Mehregany M and Matus L G 1992 SiC as a new micromechanics material *Technical Digest, IEEE Solid-State Sensor and Actuator Workshop* (*June 1992, Hilton Head, SC*) 198–201

[29] Guckel H, Randazzo T and Burns D W 1985 A simple technique for the determination of mechanical strain in thin films with applications to polysilicon *J. Appl. Phys.* **57** 1671

[30] Mehregany M, Howe R T and Senturia S D 1987 Novel microstructures for the in-situ measurement of mechanical properties of thin films *J. Appl. Phys.* **62** 3579–3584

[31] Huff M A, Senturia S D and Howe R T 1988 A thermally isolated microstructure suitable for gas sensing applications *Technical Digest, IEEE Solid-State Sensor and Actuator Workshop* (June 1988, Hilton Head, SC) 47–50

[32] Maseeh F and Senturia S D 1990 Plastic deformation of highly doped silicon *Sensors and Actuators* **A21–A23** 861–865

[33] Ding X, Ko W H and Mansour J M 1990 Residual stress and mechanical properties of boron-doped P^+ silicon films *Sensors and Actuators* **A21–A23** 866–871

[34] Ding X, Ko W H, Niu Y and He W 1990 A study on silicon diaphragm buckling *Technical Digest, IEEE Solid-State Sensor and Actuator Workshop (June 1990, Hilton Head, SC)* 128–131

[35] Ding X and Ko W H 1991 Buckling behavior of boron-doped P^+ silicon diaphragms *Technical Digest, 6th Int. Conf. Solid-State Sensors and Actuators (June 1991, San Francisco, CA)* 201–204

[36] Chu W, Mehregany M, Hasford D and Pirouz P 1992 Effect of oxidation on residual stress distribution through the thickness of p^+ silicon films *Technical Digest, IEEE Solid-State Sensor and Actuator Workshop (June 1992, Hilton Head, SC)* 90–93

[37] Wong K C 1990 *Aging Phenomena in Silicon Microcantilever Beam Light Modulator* Masters Thesis, Case Western Reserve University, Cleveland, OH

[38] Muller R S 1987 From ICs to microstructures: materials and technologies *Proc. IEEE Microrobots and Teleoperators Workshop (November 1987, Hyannis, MA)*

[39] Fujita H and Gabriel K J 1991 New opportunities for microactuators *Technical Digest, 6th Int. Conf. Solid-State Sensors and Actuators (June 1991, San Francisco, CA)* 14–20

[40] Schetky L M 1979 Shape-memory alloy *Scientific American* **241** 74

[41] Walker J A, Gabriel K J and Mehregany M 1990 Thin-film processing of TiNi shape memory alloy *Sensors and Actuators* **A21–A23** 243

[42] Johnson A D 1991 Vacuum deposited TiNi shape memory film: characterization and application in microdevices *J. Micromech. Microeng.* **1** 34

[43] Bart S F, Lober T A, Howe R T, Lang J H and Schlecht M F 1988 Design considerations for micromachined electric actuators *Sensors and Actuators* **14** 269–292

[44] Paschen F 1989 Ueber die zum Funkenübergang in Luft, Wasserstoff und Kohlensäure bei verschiedenen Drucken erforderliche Potentialdifferenz *Annalen der physik* **37** 69–96

[45] Guckel H, Skrobis K J, Christenson T R, Klein J, Han S, Choi B, Lovell E G and Chapman T W 1991 Fabrication and testing of the planar magnetic micromotor *J. Micromech. Microeng.* **1** 135–138

[46] Riethmuller W and Benecke W 1988 Thermally excited silicon microactuators *Trans. Electron Devices* **ED-35** 758

[47] Doring C, Grauer T, Marek J, Mettner M S, Trah H P, Willmann M 1992 Micromachined thermoelectrically driven cantilever structures for fluid jet deflection *Proc. IEEE Micro Electro Mechanical Systems Workshop (February 1992, Travemunde, Germany)* 12–18

[48] Suzuki K and Tanigawa H 1991 Single crystal silicon rotational micromotors *Proc. IEEE Micro Electro Mechanical Systems Workshop (February 1991, Nara, Japan)* 15–20

[49] Chen L Y, Zhang Z L, Yao J J, Thomas D C and MacDonald N C 1991 A selective CVD tungsten process for micromotors *Technical Digest, 6th Int. Conf. Solid-State Sensors and Actuators (June 1991, San Francisco, CA)* 739–742

[50] Dhuler V R, Mehregany M, Phillips S M and Lang J H 1992 A comparative study of bearing designs and operational environments for harmonic side-drive micromotors *Proc. IEEE Micro Electro Mechanical Systems Workshop (February 1992, Travemunde, Germany)* 171–176

[51] Tai Y C 1989 *IC-processed polysilicon micromechanics: technology, material, and devices* PhD Thesis, University of California, CA

[52] Mehregany M 1990 *Microfabricated Silicon Electric Mechanisms* PhD Thesis, Massachusetts Institute of Technology, Cambridge, MA

[53] Mehregany M, Bart S F, Tavrow L S, Lang J H and Senturia S D 1990 Principles of design and fabrication of variable-capacitance side-drive micromotors *J. Vac. Sci. Technol.* **A-8** 3614–3624

[54] Trimmer W and Jebens R 1989 Harmonic electrostatic motors *Sensors and Actuators* **20** 17–24

[55] Jacobsen S C, Price R H, Wood J E, Rytting T H and Rafaelof M 1989 A design overview of an eccentric-motion electrostatic microactuator *Sensors and Actuators* **20** 1–16

[56] Mehregany M, Senturia S D, Lang J H and Nagarkar P 1992 Micromotor fabrication *IEEE Trans. Electron Devices* **ED-39** 2060–2069

[57] Mehregany M and Dhuler V R 1992 Operation of electrostatic micromotors in liquid environments *J. Micromech. Microeng.* **2** 1–3

[58] Fan L S 1989 *Integrated micromachinery: moving structures on silicon chips*, University of California, Berkeley, CA

[59] Tavrow L S 1991 *A LOCOS-Based Microfabricated Radial-Gap Electric Motor* PhD Thesis, Massachusetts Institute of Technology, Cambridge, MA

[60] Tai Y C and Muller R S 1990 Frictional study of IC-processed micromotors *Sensors and Actuators* **A21–23** 180–183

[61] Bart S F, Mehregany M, Tavrow L S, Lang J H and Senturia S D 1992 Electric micromotor dynamics *Trans. Electron Devices* **39** 566–575

[62] Mehregany M, Senturia S D and Lang J H 1992 Measurement of wear in polysilicon micromotors *IEEE Trans Electron Devices* **ED-39** 1136–1143

[63] Gabriel K J, Behi F, Mahadevan R and Mehregany M 1990 *In situ* measurements of friction and wear in integrated polysilicon microstructures *Sensors and Actuators* **A21–23** 184–188

[64] Deng K and Ko W H 1992 A study of static friction between silicon and silicon compounds *J. Micromech. Microeng.* **2** 14–20

[65] Mehregany M, Nagarkar P, Senturia S D and Lang J H 1990 Operation of microfabricated harmonic and ordinary side-drive motors *Proc. IEEE Micro Electro Mechanical Systems Workshop (February 1990, Napa Valley, CA)* 1–8

[66] Mehregany M, Senturia S D and Lang J H 1990 Friction and wear in microfabricated harmonic side-drive motors *Technical Digest, IEEE Solid-State Sensor and Actuator Workshop (June 1990, Hilton Head, SC)* 17–22

[67] Lim M G, Chang J C, Schultz D P, Howe R T and White R M 1990 Polysilicon microstructures to characterize static friction *Proc. IEEE Micro Electro Mechanical Systems Workshop (February 1990, Napa Valley, CA)* 82–88

[68] Tang W C, Nguyen T H and Howe R T 1989 Laterally driven polysilicon resonant microstructures *Sensors and Actuators* **20** 25–32

[69] Brennen R A, Lim M G, Pisano A P and Chou A T 1990 Large displacement linear actuator *Technical Digest, IEEE Solid-State Sensor and Actuator Workshop (June 1990, Hilton Head, SC)* 135–139

[70] Zdeblick M J and Angell J B 1987 A microminiature electro-fluidic valve *Technical Digest, 4th Int. Conf. on Solid-State Sensors and Actuators (June 1987, Tokyo, Japan)* 827

[71] Feury A M, Poteat T L and Trimmer W S 1986 A micromachined manipulator for submicron positioning ofoptical fibers *Technical Digest, IEEE Solid-State Sensor Workshop (June 1986, Hilton Head, SC)*

[72] Moroney R M, White R M and Howe R T 1990 Ultrasonic micromotors: physics and applications *Proc. IEEE Micro Electro Mechanical Systems Workshop (February 1990, Napa Valley, CA)* 182–187

[73] Tjhen W, Tamagawa T, Ye, C P, Hsueh C C, Schiller P and Polla D L 1991 Properties of piezoelectric thin films for micromechanical devices and systems *Proc. IEEE Micro Electro Mechanical Systems Workshop (February 1991, Nara, Japan)* 114–119

[74] Senturia S D, Harris R M, Johnson B P, Kim S, Nabors K, Shulman M A and White J K 1992 A computer-aided design system for microelectromechanical systems (MEMCAD) *J. Microelectromech. Systems* **1** 3–13

[75] Crary S B and Zhang Y 1990 CAEMEMS: An integrated computer aided engineering workbench for micro-electro-mechanical systems *IEEE Micro Electro Mechanical Systems Workshop (February 1990, Napa Valley, CA)* 113–114

8

Magnetostrictive Actuators

R D Greenough, I M Reed and M P Schulze

8.1 INTRODUCTION

For several decades, the demands of sonar technology tended to direct efforts to improve the performance of transducer materials, but new technologies are now making different demands. It is not just energy conversion efficiency which is of importance—greater movement, faster speeds and higher load bearing capability are other facets of performance which have a critical rôle.

Several decades of research into piezoelectric materials have provided a firm foundation to enable their incorporation in devices for applications; the necessity for high voltages has been reduced using stacked elements, and actuation speeds remain inherently faster than metallic magnetostrictive materials. However, limitations at low frequencies and in load bearing capability are two areas which leave an ideal opportunity for magnetostrictive materials to be used. Rather than regarding the two types of materials as competitors, complementary rôles are gradually emerging and there is a renewed interest in magnetostrictive materials for actuators. Following a period of several years when the best performances could only be achieved with nickel alloys [1], larger magnetostrictive strains were discovered at low temperatures in polycrystalline rare earth elements [2] and subsequently at room temperature in binary and pseudo-binary iron-rare earth compounds [3]. A summary in table 1 illustrates the increases in magnetostriction that have been measured but the next stage, essential for future applications, was to produce these compounds in commercial quantities with sufficiently high quality. The most widely known compound, Terfenol-D, with typical composition $Tb_xDy_{(1-x)}Fe_{1.95}$ ($0.27 \leqslant x \leqslant 0.32$), is now prepared in grain-oriented form on a commercial scale by free float zoning or modified

Bridgman techniques [4] and it is at this juncture that practical applications can now be seriously considered.

Important criteria in the successful application of magnetostrictive materials include the availability of well-characterized material with repeatable performance, the prediction of response under working conditions and techniques to achieve reliable control of displacement, velocity or acceleration. This chapter addresses these issues, in particular for Terfenol-D and the practical demands which confront a device designer are analysed. Finally, applications which have been presented at seminars, conferences or published in academic journals are reviewed.

8.2 MAGNETOSTRICTIVE MATERIALS

8.2.1 Magnetic and Magnetoelastic Properties

Static field and pressure dependence. The most outstanding feature of materials often described as 'highly magnetostrictive' is the strength of the interatomic magnetoelastic coupling which causes macroscopic elastic behaviour to depend on the state of magnetization. From the viewpoint of the device designer or applications engineer, the most important parameters which collectively describe this behaviour are:

(a) magnetostriction (λ or $\mathrm{d}l/l$), the fractional increase in length, obtained by the application of a magnetic field H, which is usually an anisotropic and volume-conserving deformation;

(b) magnetostrictive strain coefficient ($\mathrm{d} = \mathrm{d}\lambda/\mathrm{d}H$), the rate at which the strain is developed with respect to the applied field; and

(c) magnetomechanical coupling (k), where k^2 is a measure of the transduction efficiency either in converting magnetic energy to an elastic deformation or vice versa.

The constitutive equations [5], which originate from Mason's representation of the behaviour of piezoelectric materials [6], define ideal relationships between magnetic and elastic properties, neglecting the hysteresis and non-linearities which occur in practice. From these, the following relationships are obtained:

$$k^2 = d^2/\mu_\sigma s_H \qquad \text{(i)}$$

$$\mu_\sigma = \mu_l/(1-k^2) \qquad \text{(ii)}$$

and

$$E_B = E_H/(1-k^2) \qquad \text{(iii)}$$

where μ_σ are μ_l are the magnetic permeabilities at constant pressure and length respectively, s_H is the compliance at constant field and E_B and E_H are the elastic moduli at constant induction and field respectively.

The relationship in equation (i) is often employed as a basis for measurement of magnetomechanical coupling, which is discussed in a later section; equations (ii) and (iii) reveal the nature of magnetoelastic coupling because the magnitude of the ordinary elastic modulus, for instance, is modified by a contribution which depends on the strength of the magnetoelastic coupling.

Each of the parameters, λ, k, d, μ_σ, E_H, as well as the magnetic induction B are both field and pressure dependent. Consequently, as figures 1 and 2 illustrate, one of the fundamental problems in the design of highly magnetostrictive devices for use in applications where the static or dynamic loading causes a change in the response of the actuator element to an applied field is that applied fields and pressures have to be coordinated so that optimum performance from the transducer can be maintained.

Hysteresis. The strain versus DC applied field curves measured under different uniaxial pressures (figure 3) show the extent of the hysteresis. Its origin has not, as yet, been identified but hysteresis appears to be greater, for instance, in $Dy_{0.7}Tb_{0.3}Fe_{1.95}$ than in $Dy_{0.73}Tb_{0.27}Fe_{1.95}$ but the magnetocrystalline anisotropy and grain orientation which provide exceptionally large d coefficients ($\approx$200 nm A^{-1}) in the former, may also be responsible for hysteresis. In the latter composition, the strain coefficient is usually smaller but the strain more linear and the hysteresis less pronounced [3]—this can be beneficial for particular applications.

8.2.2 AC Field Response

Metallic, electrically conductive magnetostrictive materials such as those listed in table 1 exhibit the same fundamental physical response to

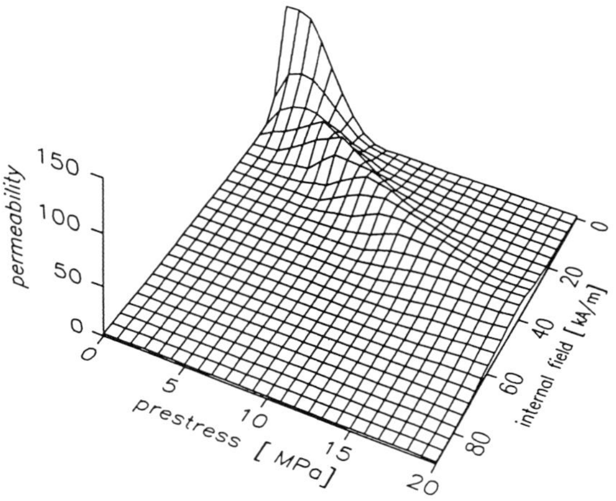

Figure 1. Field and pressure dependencies of the differential permeability, dB/dH, derived from the static B–H loops for grain-oriented Terfenol-D.

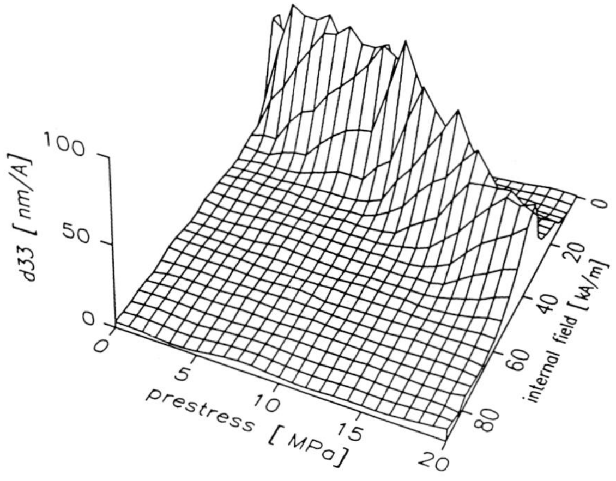

Figure 2. Dependence of the differential *d* coefficient ($d33 = d\lambda/dH$) on field and pressure derived from static $\lambda - H$ curves for grain-oriented Terfenol-D.

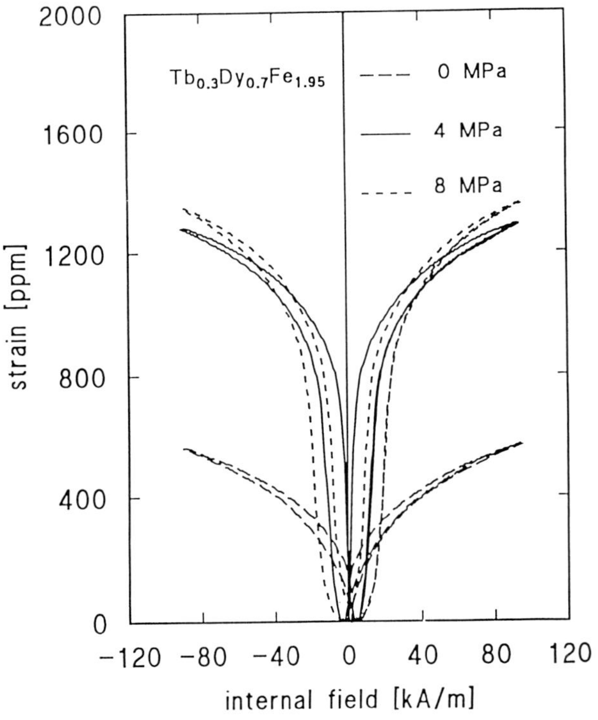

Figure 3. Magnetostrictive strain, λ, as a function of DC applied field, measured in grain-oriented material with nominal composition $Tb_{0.3}Dy_{0.7}Fe_{1.95}$, for different uniaxial prestresses applied along the $[11\bar{2}]$ grain growth axis.

Table 1. Values for the magnetostrictive strains developed in polycrystalline nickel, nickel alloys, rare earth elements and rare earth-iron compounds. (* indicates low temperature data.)

Material	λ_s (ppm)	Ref.
Ni (heat treated)	300	[1]
45 permalloy	120	[1]
2V permendur	150	[1]
Dy*	900	[2]
Tb*	840	[2]
$TbFe_2$	2700	[3]
$Tb_{0.27}Dy_{0.73}Fe_{2.0}$	1200	[3]

alternating fields; i.e. in accordance with Maxwell's equations, macroscopic eddy currents are induced which oppose the applied field. In a plane sheet, the magnetic fields and flux density decay exponentially with the depth inside the material, x, according to

$$H = H_S e^{-x/\delta} \tag{iv}$$

where H_S is the surface field strength at any particular instant of time and δ, the skin depth, is given by the equation

$$\delta = (2/\mu\sigma\omega)^{1/2} \tag{v}$$

in which μ and σ are the permeability and electrical conductivity of the material and ω is the angular frequency of the applied field.

Experimental data for the apparent permeability as a function of frequency [7] agree with this prediction (figure 4) provided excessive domain wall motion is avoided by applying a DC bias field; observations made with no bias near the origin of the magnetization curve show that the apparent permeability decreases more rapidly with frequency than estimates from skin depth calculations predict, due to eddy currents near the domain walls. The inherent frequency limit imposed by the macroscopic eddy currents can be minimized by laminating the material, in a similar manner to transformer cores and calculations by Butler [8] predict that the useful range of Terfenol might be extended to ~80 kHz using 1 mm thick laminations. Although this increases the rate of input of magnetic energy density $\frac{1}{2}\mu H^2$, the mechanical response as gauged by the corresponding elastic strain under non-resonant conditions may be slower, depending on mechanical losses and the effects of loading but a prediction that the frequency dependence of the elastic response can follow that of the input

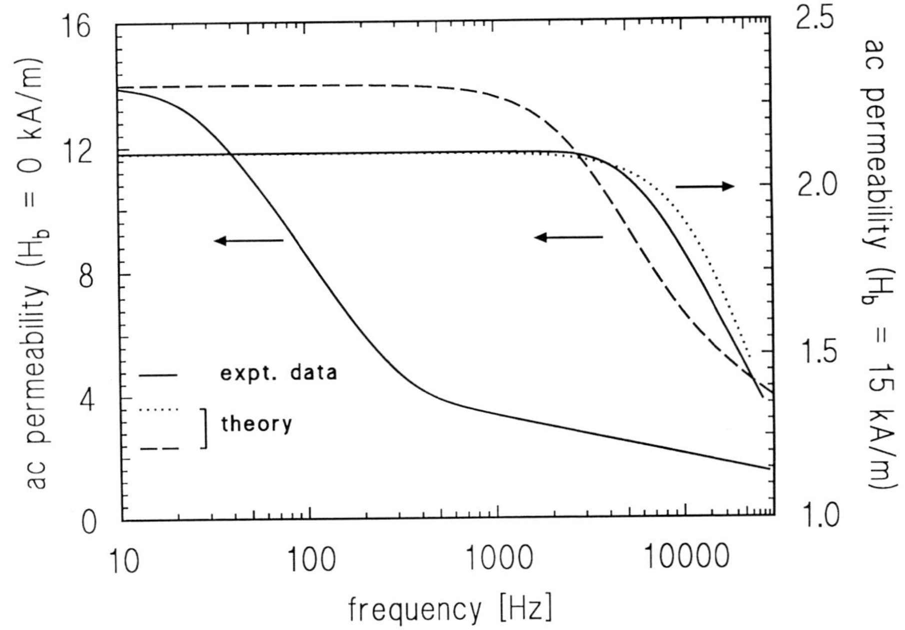

Figure 4. AC relative permeability measured (solid lines) as a function of frequency for a 6 mm diameter rod of Terfenol-D with an applied AC field of 800 A m^{-1} (p–p), with or without a DC bias field, H_b, compared with skin effect calculations for $H_b = 0$ (– – –) or $H_b = 15$ kA m^{-1} (....).

energy seems to be confirmed in practice for non-laminated materials through measurements of the frequency dependence of d_{33} [9].

The other principal factor that regulates the dynamic performance is hysteresis loss, which depends on the magnitude of the AC drive field and applied pressure as shown by studies on Terfenol by Clark *et al* [10] and Moffett *et al* [11, 12].

8.2.3 Temperature Effects

For room temperature operation, the composition of the Terfenol-D compounds is ideal but the desired performance is lost at lower and higher temperatures. As the temperature is increased, the magnetostrictive strains decrease at a rate approximately proportional to m^3 (where $m = M(T)/M(0)$, the reduced magnetization at temperature T) and $m \to 0$ as $T \to T_C$, the Curie temperature. For Terfenol-D, $T_C \sim 380$ K, the observed temperature dependencies of the magnetostriction [13] compared with theory [14], and the magnetomechanical coupling [15] are shown in figures 5 and 6 for $Tb_{0.27}Dy_{0.63}Fe_{1.95}$.

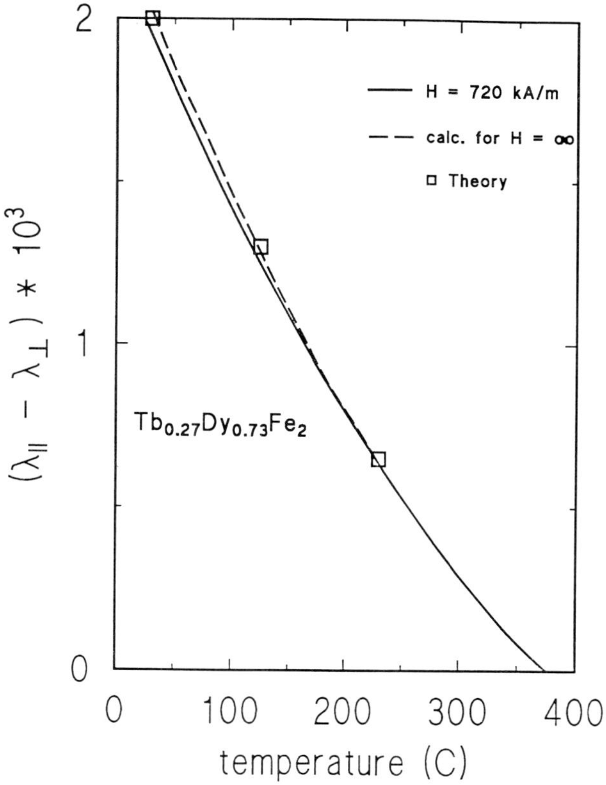

Figure 5. Temperature dependence of the magnetostriction between room temperature and the Curie temperature in a compound with nominal composition $Tb_{0.27}Dy_{0.73}Fe_{2.0}$ (solid line) compared with the theoretical single-ion behaviour predicted by Callen and Callen [14].

In Terfenol-D the anisotropies of the terbium and dysprosium ions, being of opposite sign, compete as the temperature is reduced below room temperature and consequently the easy axis of magnetization switches from a ⟨111⟩ direction to a ⟨100⟩ axis with a corresponding reduction in the magnitude of magnetostrictive strain [16, 17]. The effect occurs typically at about 273 K in Terfenol-D but this temperature is quite strongly dependent on composition.

8.2.4 Material Development

Some modifications of the properties of Terfenol-D have been considered by making changes in the composition. The goals are to raise the Curie temperature, reduce the brittleness and increase the maximum magnetostrictive strains. As well as additions of manganese, cobalt or aluminium

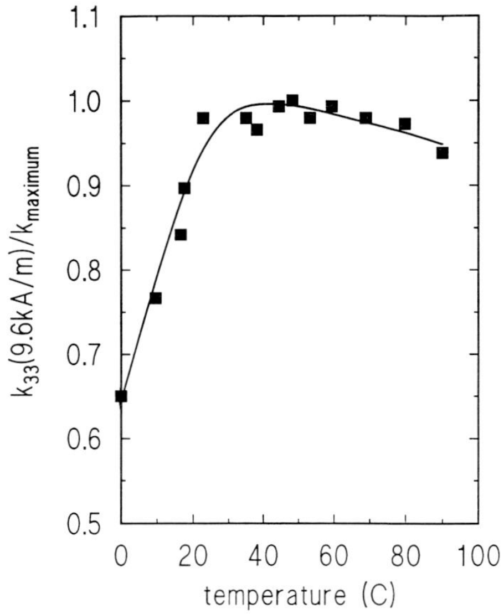

Figure 6. Magnetomechanical coupling, k_{33}, normalized to its maximum value in the temperature range $0 < T_c < 100$ °C for the nominal composition $Tb_{0.27}Dy_{0.73}Fe_{2.0}$.

[18, 19] the effects of nitrogenation have been examined in some RE_2Fe_{17} compounds [20]. Recently Tb_xDy_{1-x} compounds have been prepared [21] and have been shown to generate strains ~6000 ppm, ideally suited for use with high-temperature superconductors for incorporation in devices operating at ~77 K. A composite material consisting of Terfenol particles embedded in a non-conducting matrix [22] has advantages because it is machinable, has a higher frequency response and the magnetostrictive strains are less susceptible to applied stress, but to overcome the magnetocrystalline anisotropies associated with the random orientations of particles much larger applied fields are required compared with grain-oriented Terfenol-D.

8.3 CHARACTERIZATION AND NON-DESTRUCTIVE TESTING OF MATERIALS

8.3.1 Characterization

The constitutive equations which lead to equations (i)–(iii) form the basis of present techniques to characterize magnetostrictive materials. Data for λ, d, μ and k can be obtained under either DC or AC field conditions and although DC field data may not be totally representative of materials performance under dynamic working conditions, they are sensitive to structural imperfections and inhomogeneities in Terfenol [23] and hence act as a gauge of material quality. Furthermore, DC field data can be used as a figure of merit for device performance but which particular parameter is of greatest significance depends on the ultimate application. For sonar, where overall power dissipation is not of prime importance the magnetomechanical coupling is chosen; for actuators in machines where power and physical size are significant factors in the design, the *d* coefficient is more important than the coupling in order that small applied fields can generate large displacements.

There are two principal experimental methods to measure magnetomechanical coupling as a function of applied field and uniaxial pressure [24] which use either equation (i), measuring each of the parameters d, μ_σ and s_H in three separate experiments, or equation (iii) in which the elastic moduli are measured from sound velocities under isofield or constant induction conditions. These velocities can be obtained by monitoring the longitudinal resonant response of a cylindrical rod, measuring the resonant and antiresonant frequencies f_R and f_A respectively. Hence the magnetomechanical coupling can be calculated from the expression

$$k_{33}^2 = \pi^2/8(1 - (f_R/f_A)^2). \qquad \text{(vi)}$$

A comparison of these methods [25] attempted to show that provided the first method incorporated measurements made under identical conditions (i.e. differentiation of static B–H and λ–H curves to obtain μ_σ and d_{33} under the same pressure and field conditions or the same conditions for their incremental values) the results for k_{33} were in close agreement with a second method using 'dynamic' resonance measurements. However, even though AC field response has been corrected for the skin effect [26] this agreement may be fortuitous as shown by a more recent, rigorous analysis [27].

Techniques have now been developed which enable the measurement of B–H and λ–H curves, k_{33}, d_{33} and μ_σ, as functions of applied pressures up to 50 MPa, with DC bias fields ≤ 140 kA m^{-1} and AC fields ≤ 500 A m^{-1} using the equipment shown in figure 7. The samples are subject to non-uniform demagnetizing fields but no simple solution to this problem is available; even the use of an electromagnet with samples in close proximity to or in contact with the magnet poles cannot avoid the demagnetizing field problems unless the permeabilities of the sample and pole pieces are always matched at every part on the magnetization curve.

8.3.2 Non-destructive Testing

Results from the characterization of different samples of Terfenol over a period of several years show how the quality of commercially available

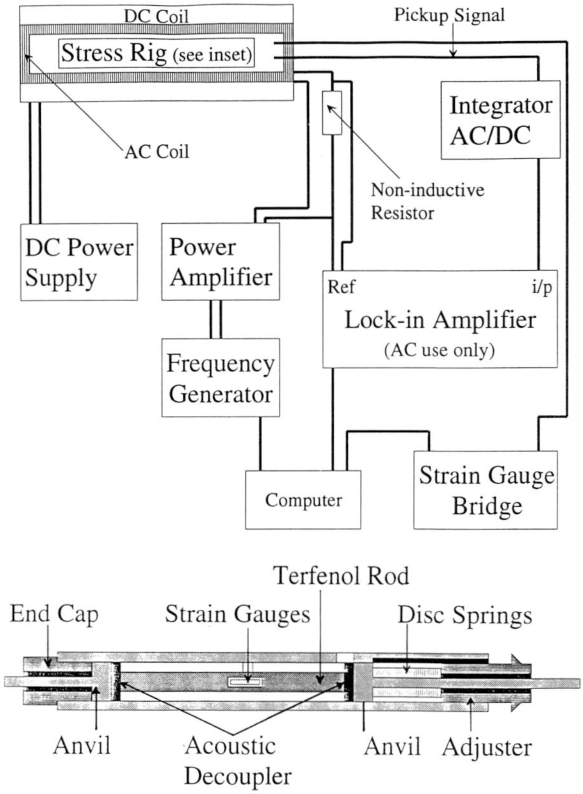

Figure 7. Experimental equipment to characterize magnetostrictive materials by measuring, for instance, DC field magnetostrictive strain and magnetic induction together with their field derivatives, AC incremental permeability, *d* coefficient and magnetomechanical coupling as a function of field strength and applied uniaxial prestress ≤50 MPa.

material has improved (figure 8), attributable to better grain orientation, chemical composition and homogeneity [28, 29, 30]. However as a check on quality, a non-destructive testing technique has been developed which can detect these imperfections as well as pores, voids or cracks [31]. It measures either the AC permeability or static induction B at points along the length of a cylindrically shaped sample as a function of DC bias field applied along the sample axis. Typical results shown in figure 9 are from scans along rods of good and poor quality Terfenol-D and the principal features in the latter—lack of symmetry and fluctuations—have been correlated with chemical gradients and grain misalignments respectively [23].

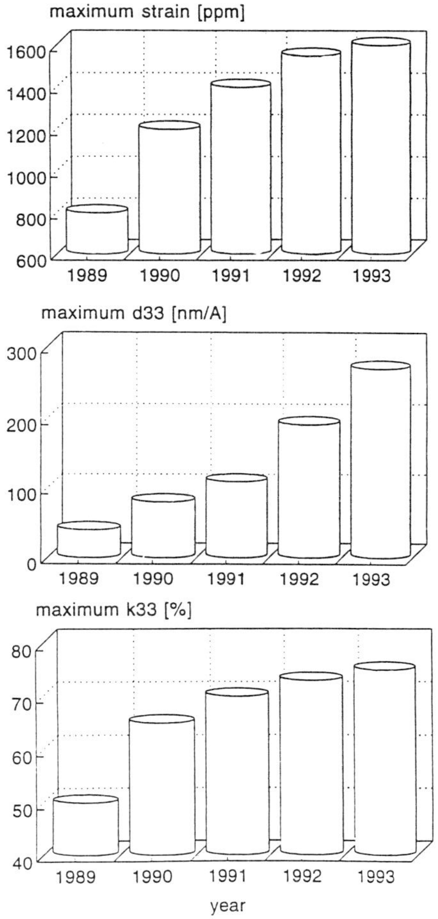

Figure 8. Recent improvements in the performance of Terfenol-D, seen through the characterization of many samples of the same nominal composition, according to the maximum values obtained for the static magnetostrictive strain (in a field of 140 $kA\,m^{-1}$), d coefficient and coupling efficiency.

8.3.3 Impedance Analysis

Following the work of Butterworth and Smith in 1930 [32], a method employed either to characterize magnetostrictive materials or assess device performance is to measure the real and imaginary components of electrical impedance as a function of frequency and analyse the motional impedance loop (the plot of real versus imaginary impedances) formed near the magnetomechanical resonance. This can simply provide values for f_R and f_A

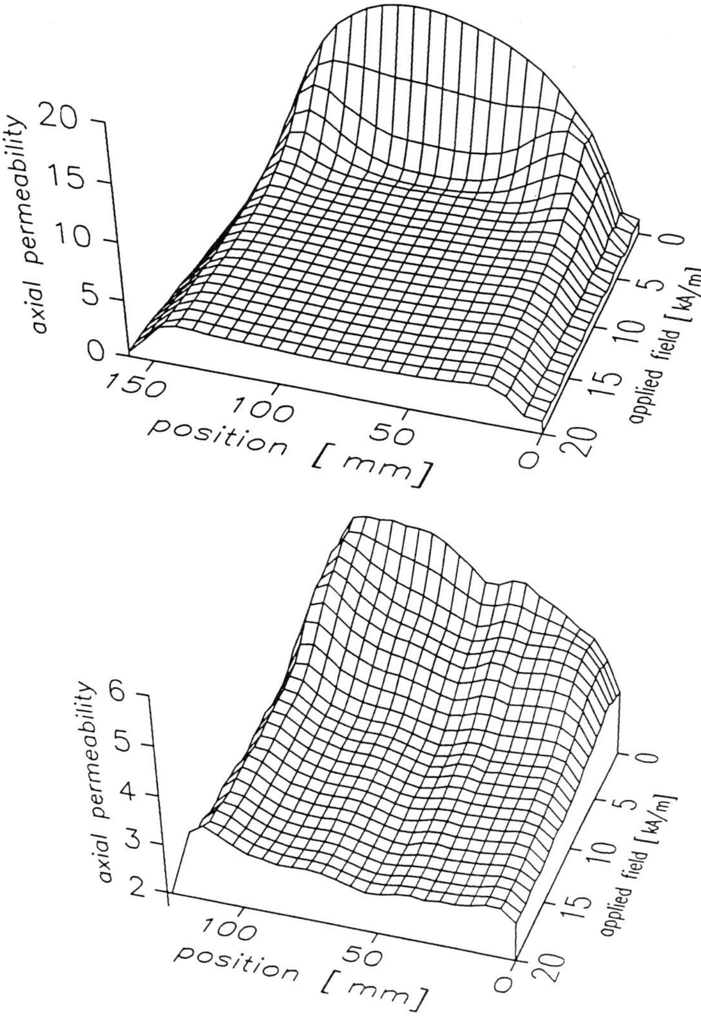

Figure 9. Non-destructive testing using AC permeability scans along the length of two grain-oriented Terfenol-D rods, for a range of DC bias fields. The upper scans were obtained from a sample with acceptable grain orientation and chemical homogeneity; the lower scans, from a sample of relatively poor quality, reveal grain misalignments and composition gradients.

to obtain coupling efficiency and quality factor Q [33] but for a more sophisticated analysis, plane wave modelling (PWM) can be employed, which takes eddy current, hysteresis and mechanical losses into account, together with their frequency dependencies which are responsible for acircular impedance loops. The PWM, first proposed by Earwicker [34] and developed by Claeyssen [35] is a computer simulation technique which can be employed to predict a device output according to its construction or, conversely, yield information on the characteristics of a magnetostrictive drive rod inside a device from an analysis of the observed device impedance [36]. The one-dimensional wave equation

$$dT/dx = \rho d^2u/dt^2 \qquad \text{(vii)}$$

where T is the stress, ρ the density and u the displacement, is coupled with the constitutive equations and solved according to the boundary conditions. Losses are included by incorporating complex values for the permeability and compliance.

In a recent paper [27] the benefits of PWM have been demonstrated in an analysis of experimental electrical impedance data obtained from a 'free standing' rod of Terfenol-D. The magnetostriction parameters obtained from an iterative fitting of the model to the real and imaginary components of the impedance data from a free standing rod of Terfenol (figure 10) are shown in table 2. The reliability of the method, with error analysis, was tested on simulated resonance data using assumed values for the magnetostriction parameters. It is found that the values of k_{33} derived by the PWM method in either the simulated or experimental case are different from those obtained using the values of just f_R and f_A in equation (vi) because PWM accommodates the loss mechanisms during the analysis instead of attempting a retrospective correction to a 'loss free' result.

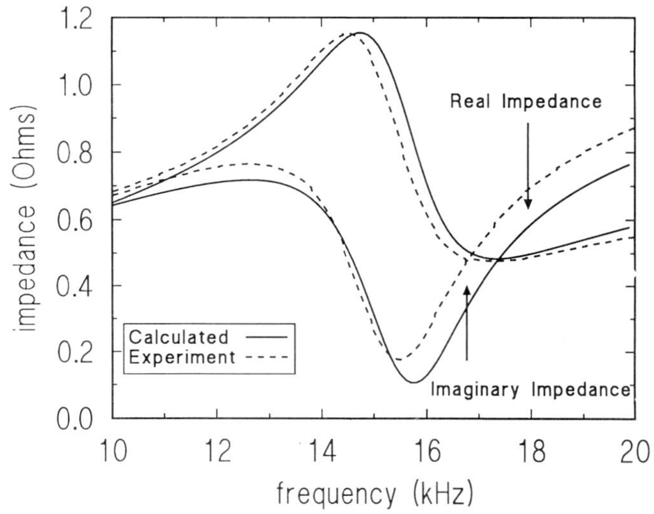

Figure 10. Electrical impedance as a function of frequency measured in a free-standing rod of Terfenol-D compared with the impedance calculated from the plane wave model (derived parameters are given in table 2).

Table 2. Material parameters for a Terfenol-D sample derived from a fit of the plane wave model (PWM) to the experimental impedance data shown in figure 10 (for a bias field of 25 kA m^{-1}). The imaginary components of the permeability and elastic compliance are represented by $\mu_{33}^{S''}$ and $s_{33}^{B''}$ respectively, the real components by $\mu_{33}^{S'}$ and $s_{33}^{B'}$.

Parameter	Units	Value
g_{33}	T^{-1}	$(1.114 \pm 0.003) \times 10^{-3}$
$\mu_{33}^{S'}$	H m^{-1}	$(3.480 \pm 0.008) \times 10^{-6}$
$\mu_{33}^{S''}$	H m^{-1}	$(-1.188 \pm 0.011) \times 10^{-7}$
$s_{33}^{B'}$	m^2 N^{-1}	$(1.663 \pm 0.001) \times 10^{-11}$
$s_{33}^{B''}$	m^2 N^{-1}	$(-1.037 \pm 0.007) \times 10^{-13}$
k_{33}	—	0.509 ± 0.002

8.4 DEVICE DESIGN, PERFORMANCE AND CONTROL

8.4.1 Actuation Principles

The task that faces the device designer is to specify the construction of an actuator which meets the requirements of a particular application. As the observed pressure and field data for Terfenol show (figures 2 and 3) the

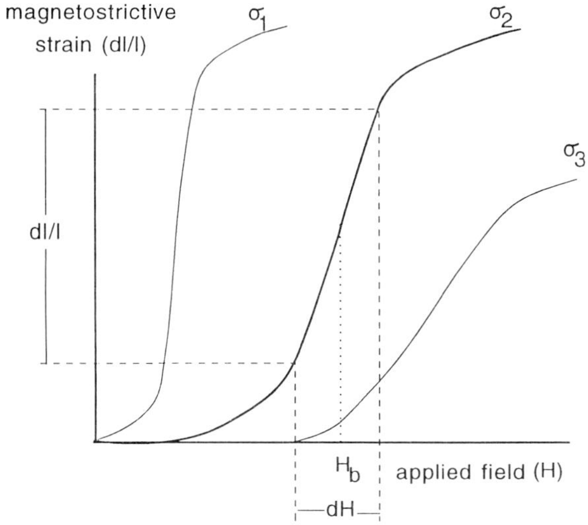

Figure 11. Magnetostrictive strain generated by the application of a quasi-static or oscillating field, ΔH, superimposed on a DC bias field H_b for one of three different uniaxial pressures, σ_1, σ_2 and σ_3.

optimum strain and strain coefficient (and consequently, coupling efficiency) cannot be obtained simultaneously with the same bias field and pressure. The schematic diagram in figure 11 illustrates the task more directly. Let us suppose that a mean operating point is chosen by applying a DC bias field, H_b, and that the strain limits are obtained from application of a quasi-static or oscillating drive field of magnitude ΔH. If the load on an actuator varies during the course of operation, then to obtain constant displacement output, both the bias point and drive field have to be altered accordingly by an amount which will depend on the demands of a specific application. Even this is an oversimplified representation because of the presence of hysteresis and non-linearities, both of which depend on the pressure and on the magnitude and frequency of the AC drive field, adding to the demands on an ideal control system.

Table 3. Maximum strains (ppm) measured under the specified applied fields and uniaxial stresses measured in two compositions of Terfenol-D prepared by a modified Bridgman process (Etrema Products Inc., USA); their means and standard deviations.

Stoichiometry $Tb_{0.285}Dy_{0.715}Fe_{1.9-1.95}$ Field: 1000 Oe Prestress: 10.2 MPa (1.5 Kpsi) Diameter 35 mm (1.2 in) Length 25 mm (1.0 in)			Stoichiometry $Tb_{0.3}Dy_{0.7}Fe_{1.9-1.95}$ Field: 1000 Oe Prestress: 13 MPa (1.9 Kpsi) Diameter 15 mm (0.6 in) Length 76 mm (3.0 in)		
Production ID	Mean	Std dev	Production ID	Mean	Std dev
BRDG-1-037	1198	57	BRDG-2-023	1303	9
BRDG-1-039	1151	65	BRDG-2-025	1293	5
BRDG-1-041	1164	59	BRDG-2-027	1299	34
BRDG-1-043	1256	61	BRDG-2-021	1290	0
BRDG-1-045	1217	31	BRDG-2-033	1300	0
BRDG-1-059	1284	42	BRDG-2-035	1247	33
BRDG-1-065	1253	67	BRDG-2-037	1278	41
BRDG-1-093	1243	65	BRDG-2-039	1220	20
BRDG-1-099	1288	34	BRDG-2-045	1330	16
BRDG-1-101	1190	84	BRDG-2-047	1253	13
BRDG-1-103	1265	44	BRDG-2-057	1332	37
			BRDG-2-059	1298	19
			BRDG-2-079	1267	40
			BRDG-2-081	1242	66
			BRDG-2-090	1295	23
GROUP	1228	73	GROUP	1266	42

An additional complication will arise when there is a lack of reproducible behaviour from one sample of Terfenol to another, even though they may have the same nominal composition. The reproducibility of material produced by Etrema Products [37, 38] is shown in table 3 as gauged by the magnitude of the magnetostrain obtained under the specified conditions: 1000 Oe (~80 kA m^{-1}) applied field, a uniaxial applied stress of 10.2 MPa for the composition $Tb_{0.285}Dy_{0.715}Fe_{1.9-1.95}$ and a prestress of 13 MPa for the composition $Tb_{0.3}Dy_{0.7}Fe_{1.90-1.95}$. These strain levels give only one indication of reproducibility and other parameters may be more significant for device usage, such as the *d* coefficient which tends to be more sensitive than strain to grain misalignment or chemical inhomogeneity [23]. Material production techniques have been developed and quality improved during the last three or four years but in common with piezoelectric materials, exactly reproducible performance from one sample to another to the extent that one device can be replaced by another with identical behaviour remains a long term goal. In the mean time, the user can build a device based on non-destructive testing and characterization data (as described in section 3) and depending on how stringent the specifications are which have to be met, use a self-adaptive control system to obtain precise, predetermined device output.

8.4.2 Device Components

In a typical magnetostrictive actuator, the bias field H_b can be provided by permanent magnets, and to reduce total power consumption and minimize heat generation, only the actuation power is supplied by a solenoid. Flux closure is necessary for maximum efficiency. Optimum prestress, especially for low load applications, can be provided by springs with a selected compliance and time constant.

Some of these fundamental features have been studied in detail using finite element analysis to investigate magnetic flux and stress distributions for instance. In a study of magnetic biasing techniques by Sewell and Kuhn [39] and Vermet *et al* [40], different configurations of permanent magnets and flux closure systems have been compared to determine the arrangement which generates the most uniform and highest flux density along a Terfenol-D actuation rod. Difficulties arise due to the low relative permeability of Terfenol-D ($\mu/\mu_0 \sim 5$) just when the *d* coefficient is at a maximum, and to the changes induced in its reluctance due to the application of pressure. Stähl *et al* [41] analysed the pressure distribution in loaded Terfenol rods and demonstrated that stress concentrations occur near the edges of an actuator rod, especially in regions where the rod interfaces with a different material in the assembly. Several finite element packages are available commercially for these flux and stress analyses.

8.4.3 Resonant and Non-resonant Actuators

The frequency response of an actuator is determined not only by the inherent speed of the magnetostrictive material but is strongly influenced by the mechanics of the system—the physical construction of the device and its Q factor. The reaction of the load also has to be considered: a simple example is encountered when measurements are made on a free standing Terfenol rod in resonance to measure coupling efficiency. The rod has to be acoustically decoupled from the pressure anvils (cf figure 7), otherwise motion in the sample is so distorted by resonances in the mechanics of the system that the rod's resonant and antiresonant frequencies cannot be found. On the other hand, the resonant response from the mechanical housing of a transducer can be used to good effect by amplifying the vibration. Claeyssen reports a doubling of the maximum vibration amplitude in a sonar device [42].

An actuator which is to perform in a non-resonant mode is more difficult to design, and its speed is mainly governed by the inherent properties of the magnetostrictive drive, inductance of the transducer together with the damping and inertial effects of the whole device and its load.

8.4.4 Device Modelling

Some modelling methods and techniques have already been considered in the context of material characterization. A free standing rod of Terfenol is a particularly simple example of a device when the ancillary equipment such as solenoids, sensors etc are included as part of the assembly. Such work was directed towards characterization of magnetostrictive material, but other models have been developed which are designed specifically to predict output from an integral device, such as a sonar transducer. One of the oldest is the equivalent circuit (lumped parameter) method in which magnetic and mechanical components are represented by electrical component analogues. The surroundings of a device can be included in the model and hence its response in a working environment, such as water, can be predicted. The technique has been used by Butler [43] for the particular case of a Terfenol transducer. In the equivalent circuit, shown in figure 12, M and m represent

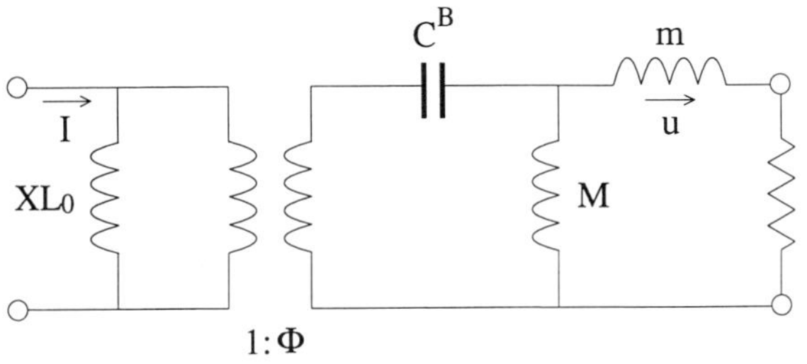

Figure 12. Lumped parameter representation of a magnetostrictive device.

the masses at each end of the device, the former remaining stationary and the latter with velocity u if $M \gg m$; R is the mechanical damping associated with mass m. The inductance of the clamped (i.e. constant length, non-resonant condition) Terfenol rod is represented by L_o and the transducer's overall magnetomechanical coupling, which includes the inherent coupling of the transducer rod and the mechanics of the device, is represented by the transformer ratio, Φ. Compliance of the spring loading between the head mass, m, and actuation rod is represented by C_2 (the reciprocal of the spring constant k_2), and C_B is the short circuit Terfenol rod elastic compliance (S_B) measured with constant magnetic induction. Further structures can be connected to the assembly by adding 'T' networks and load on the head mass can be a complex quantity to take into account the radiation conditions (e.g. radiation into air or water for instance). Here the overall impedance of the transducer as a function of frequency, especially the resonant response, can be predicted to find head mass displacements or velocities. An equivalent circuit representation has also been employed by Kvarnsjö and Svensson [44] to model the effects of eddy currents and hysteresis losses. However, the equivalent circuit model has some limitations. It is rather inflexible when variations in load occur, static or dynamic, because the resultant pressure changes on the Terfenol rod then cause its magnetomechanical coupling to vary. Such effects have to be accommodated by changes in the transfer ratio Φ, which then has to be treated as a variable which may have to represent quite rapid changes in load. Furthermore, equivalent circuits cannot simulate devices at operating frequencies much greater than the principal resonance frequency.

An alternative simulation proposed by Kvarnsjö and Engdahl [45] employs DC magnetostriction data to predict the dynamic performance of a transducer rod by dividing it into a finite number of sections. With strains x_i, stress σ_i and elastic modulus E_i in the ith section, assuming an elastic response, Newton's Law is applied in the form

$$\sigma_{i-1} - \sigma_i = \rho d^2 \frac{\mathrm{d}^2 x_i}{\mathrm{d}t^2} + 2d\sqrt{E_i \rho \xi}\, \frac{\mathrm{d}x_i}{\mathrm{d}t} \qquad \text{(viii)}$$

where ρ is the density and d a section length; the damping coefficient ξ is used as an adjustable parameter to regulate the extent of displacement overshoot or oscillatory motion. Outputs for different load masses are predicted. This model was subsequently extended to include the effects of macroscopic eddy currents together with hysteresis losses represented by the Priesach model [46]. These analytical elements were then combined in a computed electromechanical transfer function for a device [47] using finite-element flux distribution and variable levels of DC flux reluctance, hence leading to the frequency dependence of the transfer function.

A rewarding method to predict transducer performance (or, used in reverse, to determine the properties of a magnetostrictive transducer element at the heart of the transducer) is the plane wave model (PWM), described in section three. Through the use of matrices containing wave propagation information, the acoustic coupling between the separate components of a device can be incorporated. Each component is attached to another in one of two ways, in series or in parallel, both having an associated matrix transformation which can ultimately lead to the modelling of not just the magnetostrictive material, but the entire device. The method still requires development to cater for response in three dimensions.

8.4.5 Actuator Control

It is beyond the scope of the present work to provide an extensive analysis of work conducted on actuator control but often the need for control is important and worth consideration. The lack of exact reproducibility and the presence of hysteresis in the transducer materials together with their relative nature i.e. change in magnetic properties with applied load—usually demand a control system for applications such as accurate positioning or active vibration control. The basic principles of feedback control can be found in connection with piezoelectric, hydraulic or motor control as well as for magnetostrictive devices. For instance, Daley [48] reports the use of a PC-based self-tuning controller applied to an electrohydraulic system and having observed that a PID control becomes unstable under particular operating conditions, introduced a self-tuning controller with a sample time of 12 ms, and off-time processing of approximately 300 data points. Response to changing load is investigated by Raab and Iserman [49] for an adaptive parameter controller and positioning accuracy of 15–20 μm was obtained from an electropneumatic actuator.

A control strategy which combines variable structure and adaptive self-tuning techniques is described by Greenough *et al* [50]; the self-tuning property ensures that the variable structure strategy is operated along an optimized feedback switch line i.e. between positive and negative feedback control [51]. This system is specifically designed to operate with Terfenol as an actuator for active vibration control.

8.5 MAGNETOSTRICTIVE DEVICES AND APPLICATIONS

8.5.1 Introduction

The present section is concerned with the practicalities of actuator construction and applications. As far as the latter are concerned, the present study is not intended to be either an exhaustive in-depth study or a simple listing of all reported applications for highly magnetostrictive devices. For convenience, applications have been grouped into four main categories—*sonar*, *mechanical actuators*, *shock and vibration isolation* and *miscellaneous*, the last category containing a variety of applications which individually make a notable contribution to the field.

To some extent, highly magnetostrictive materials have been in a situation similar to that of the laser many years ago described then as 'a solution looking for a problem'. However, several novel prototype devices have emerged in recent years which are commercially viable. An up-to-date comprehensive review is restricted by work which is withheld for commercial reasons and many new patents will emerge in due course in which Terfenol technology is tailored to meet the demands of specific applications.

One misconception needs rectifying: highly magnetostrictive materials in devices do not generate large displacements in the absolute sense. Hydraulic

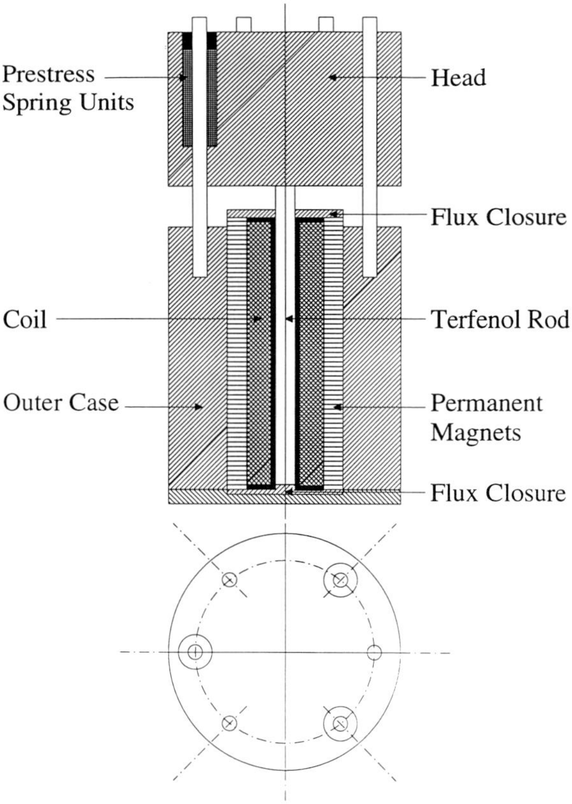

Figure 13. Principal components of a magnetostrictive actuator designed to work under optimum prestress and field bias with minimum power requirements.

pistons or conventional solenoid actuators can provide much larger strokes, but Terfenol provides exceptional force for controlled, directly driven displacement. For example, movement from a cylindrical bar of Terfenol ~250 mm long can be seen without any amplification and with a 20 mm diameter creates forces in excess of 3.5 kN. For many engineering applications though, useful displacements of several millimetres or more are required and a suitably compact arrangement has to be made to amplify the movement through levers or hydraulic systems for instance. The force or load bearing capability of Terfenol makes this possible so that usable power levels are still available with the amplified stroke.

8.5.2 Basic Terfenol Device

The construction of a general purpose actuator designed in accordance with the criteria outlined in the previous sections is shown in figure 13. Typical DC field performances for two actuators are shown in figure 14 and the output capabilities of the different sizes of actuator are given in table 4. In this case the components for the devices have not been designed or selected to meet the specifications for any particular application but designed to give optimum performance under constant load conditions.

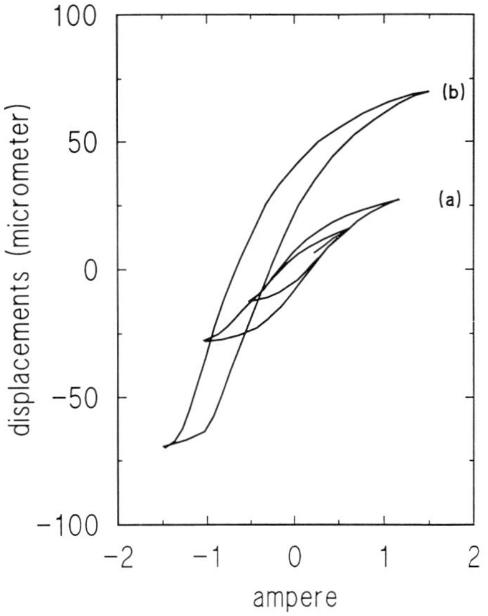

Figure 14. Displacement versus current for two actuator designs, each with permanent magnet biasing. (a) Etrema Products Inc. (USA) type 50/6-M, rated load 490 N, coil resistance ~6.4 Ω. (b) Magnetostrictive Technology Systems, Hull University (UK), rated load 1kN, coil resistance ~10 Ω.

Table 4. Typical performance figures for several Terfenol-D actuators.

			Sizes (mm)	
Stroke (μm)	Load (kg)	Power req. (W)	Length	Diameter
250	350	30	300	14
200	28	30	150	50
50	25	16	64	25

The AC response for a small device (50/6—MP from Etrema Products, USA) shows a frequency response in which the measured acceleration remains almost constant in the range $1 \leqslant f \leqslant 20$ kHz. Measurements of response times measured from the displacement of an uncontrolled, unloaded actuator constructed with a 210 mm long, 6 mm diameter Terfenol rod are shown in figure 15 relative to the drive current; rise times of ~2 ms in mechanical response are observed, an indication of the typical speed of a non-resonant actuator.

8.5.3 Sonar Transducers

These are mainly resonant devices and the total response of the Terfenol

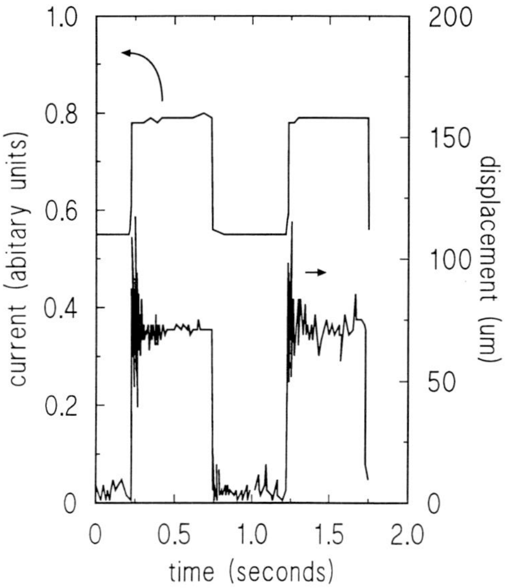

Figure 15. Response of an uncontrolled, undamped actuator to a square wave current input with displacement rise time ~2 ms.

drive element and housing determine the final output; the magnetostrictive material's performance is only one contributory factor to an overall magnetomechanical response that is used to gauge device performance.

The first designs for underwater transducers incorporating Terfenol appeared as early as 1977. A folded horn transducer was designed and constructed by Smith and Logan [52] to be as similar as possible to an existing piezoelectric transducer for comparison purposes. It contained two actuation rods, 2.4″ long and 0.25″ in diameter. Both static and dynamic fields were generated by a single solenoid. The DC field ~80 kA m^{-1} was generated with a current of ~6 A, and ~2.1 A was employed to cause the excitation. The device was multiresonant, like its piezoelectric counterpart, due to the mechanical effects of the horns and stress bolts. The elastometer seal between the horns was the origin of the device's real component of impedance which increased with frequency, and the whole transducer performance was reported to be very sensitive to the type of seal used. In the absence of information necessary to optimize the performance of the Terfenol under working conditions, the appropriate level of stress biasing was not applied, which was considered to be one reason why the output of this device was ~20 dB less than its piezoelectric counterpart. Another reason for lower output was that the volume of magnetostrictive material employed was more than 40 times less than the volume of piezoelectric. In the light of these observations, it was considered that with improved mechanical design, the magnetostrictive unit would then generate much greater output levels, especially at low frequencies, and that with permanent magnets to provide DC bias fields, power consumption would be considerably reduced.

Meeks and Timme [53] reported results of a low-frequency (~2.6 kHz) Tonpilz, longitudinal, unidirectional, piston transducer, again constructed with Terfenol to compare with a piezoelectric transducer of the same dimensions. The greatest loss in efficiency was due to eddy current loss in the actuation core so that the incorporation of laminated rare earth–iron rods was recommended. The advantages of Terfenol compared with the piezoelectric were again seen to be at low frequencies.

In either of these two studies of sonar transducers, no indication of Terfenol quality was given but at about the same time in a bi-directional device described by Akervold *et al* [54], the Terfenol was arc cast material with a maximum differential d coefficient ~11 nm A^{-1}, well below the level that can be expected from present day grain-oriented material. These initial designs, in spite of their shortcomings, are a useful guide to the design of more efficient, powerful devices and have encouraged further work in this area.

The design and performance of a resonant octagonal transducer was published in 1980 by Butler and Ciosek [55], with a circumferential resonance at ~775 Hz and mechanical quality factor, Q, of ~2.4. Both DC and AC fields were again generated with current. A DC field of ~19.2 kA m^{-1} was obtained from a coil around each of the elements with ~1000 turns of 26 gauge wire. Prestress of ~14 MPa (~2 Kpsi) was applied to each of the Terfenol rods. The effective overall magnetomechanical coupling was found to be ~0.4, about 27% less than the quasi-static coupling measured in the 'free standing' drive rods and this was attributed to both an inadequate prestress mechanism and the stiffness of the polyurethane encapsulation. At the resonant frequency, eddy currents were found to be negligible and losses were mainly mechanical and ohmic rather than in the Terfenol. Output was measured at 175 dB re 1 μPa @ 1 m per ampere of AC drive when operated at highest power efficiency. Even then the Terfenol was below saturation and it was estimated that the power density could therefore be increased by ~50% for this actuator configuration.

Oswin *et al* [56] investigated magnetostrictively driven piston and flextensional transducers using Terfenol-D with a magnetomechanical coupling ~0.65, producing a typical maximum output of ~168 dB re 1 μPa @ 1 m per ampere. Recently data have been available from commercial transducers which specify levels of ~207 dB re 1 μPa @ 1 m at a frequency of 550 Hz [57].

The most recent design study and construction exercise has been undertaken by Claeyssen [42]. The Terfenol was characterized *in situ* through a PWM analysis of the device's performance, from which the particular sample of Terfenol was observed to have a maximum magnetomechanical coupling coefficient of ~0.7 with an applied pressure of 33.1 MPa. A significant feature was the amplitude of vibration measured from the device at resonance when dynamic strains ~2500 ppm far exceeded the maximum static strain levels of ~1600 ppm. Compared with a piezoelectric transducer, sound levels of ~216 dB re 1 μPa @ 1 m in the magnetostrictive device were ~12 dB higher. The relationship between power handling and Q factor for Terfenol or PZT4, also investigated by Moffett *et al* [11], demonstrated the superior performance of Terfenol especially,but not exclusively, in low-Q devices.

The largest Terfenol device reported to date is an ocean tomography transducer designed to operate at 250 Hz and depths to about 1000 metres [58].

8.5.4 Mechanical Actuators

Another application which has attracted attention is the general use of Terfenol to actuate mechanical systems. Savage *et al* [59] estimated that at 1 kHz a power density figure of merit would be ~3×10^7 W m^{-3} with a drive field of 40 kA m^{-1}, well above the conventional figure of merit for conventional hydraulics. Conventional hydraulics exhibit a figure of merit 10 to 20 times that of electric motors. Prototypes of linear and rotary motors have now been constructed.

The elastic wave motor (EWM) is a linear actuator which combines precise movements with high force. It is often referred to as the 'Keisewetter' motor after its inventor and has potential for applications in manufacturing processes, valve actuation or machine tool control. It consists of a Terfenol rod held firmly by friction inside a tube (the stator) which exerts a radial stress on the rod. The stator carries a series of coils along its length which are powered in sequence by bursts of current. At the position of an energized coil, the Terfenol expands in the longitudinal direction (length change δl) and simultaneously contracts radially (the magnetostriction is a volume conserving strain), so releasing itself from the stator. When the current is reduced to zero in that coil, this section of rod is reclamped and the adjacent coil is energized; the process is continued until the field coil at the end of the rod is finally powered and the sum of all the individual displacements then appears at that end. Repetition of this process can produce strokes of up to 200 mm, a step size ~10 to 20 mm s^{-1} depending on how many coils are energized simultaneously. With ~80% of the rod quiescent at any instant, loads greater than 1 kN can be sustained but this does depend on the machining tolerances of the Terfenol rod and tube—typically needing ⩽10 μm surface smoothness for a small diameter rod—and on the surface wear of the rod and tube. The motor has the advantage that the rod is immobile when the current is zero. For further details of the EWM and a critical appraisal of the device, the reader is referred to a report by Roth [60].

The construction and test results of a direct drive rotary magnetostrictive motor have been published by Vranish *et al* [61] incorporating the compound $Tb_{0.3}Dy_{0.7}Fe_2$. Like the linear version, it operates on the 'inch worm' principle, incorporating two drive elements powered sequentially to rotate in steps of ~800 microradians, clockwise or anticlockwise. The prototype produced 12.2 N m of torque direct from the drive shaft at a speed of 0.5 rpm, consuming 600 W. The high torques are seen to be of value in the aerospace and robotics industries where present technology suffers from gearbox backlash and excessive weight. The paper by Vranish not only presents information on the prototype's performance, but a description of the design and expected performance figures for torque and frequency.

Exhaust and fuel injection valves actuated magnetostrictively have been considered by Dyberg [62]. A lever mechanism with a ratio ~118 was used to obtain 10mm of movement at 50 Hz for an exhaust value; and a ratio of 3.4 generated lift-off ~400 μm at 12 Hz against a fuel injection pressure of 600 bar. Fields ~160 kA m^{-1} were used to generate the magnetostriction in Terfenol rods 100 mm long and ~10 mm in diameter, working against a minimum spring pressure of ~7 MPa.

Novel magnetostrictive actuator concepts with long strokes for application to brakes, valves or remote control systems have been considered by Adolphson [63]. 'Wedge', 'thread' or 'grip' actuators are described. Speeds of 60 mm s^{-1} moving a mass of 1 kg were obtained with the grip actuator and strokes of ~100 mm at a speed of 100 mm s^{-1} with the thread actuator.

8.5.5 Shock and Vibration Isolation

Conventional passive 'spring and damper' vibration isolation systems have some distinct drawbacks.

(i) For effective low-frequency isolation, the mass of a table, for instance, needs to be extremely large.
(ii) Transmission through the spring and damper becomes greater at low frequencies. To combat this, the mount has to be made soft, but difficulties are then encountered in maintaining a steady static level. Relaxation times become long and restoration of a table level may take several seconds. This problem is encountered if the loading changes.

The concept of active vibration cancellation is attractive because these problems can be avoided. Practical solutions were considered long before the advent of magnetostrictive actuation. The first commercial systems, marketed several years ago, combined passive and active mounts to operate at frequencies between 1 and 100 Hz. Outside these limits the performance is similar to a passive isolator with a resonance at ~10 to 20 Hz. Servo-systems regulated the horizontal and vertical movements with 'voice coil' actuators which limited the dynamic loading but now, magnetostrictive transducers overcome these load or force limitations and will operate at very low frequencies, even with steady (DC) fields. There are the attendant problems associated with the inherent static and dynamic properties of the magnetostrictive materials which have already been addressed, and it is usually necessary that a feedback control system is employed with suitable transducers to sense acceleration, velocity or displacement.

Experimental and theoretical work on active shock isolation employing a magnetostrictive actuator has been described by Reed [64]. The magnetostrictive actuator was soft mounted on a base and carried a platform to which sensors were attached to measure platform displacement and its first two derivatives. The base was disturbed by a single 1 kHz, 100G pulse and its effect monitored for comparison with the predicted response of the magnetostrictive device functioning from an analogue controller. With sufficient acceleration gain to have some influence on the output frequency response, instabilities rendered the equipment unstable to unwanted external disturbances, a problem attributed in part to the non-linearity of the magnetostrictive actuator. A solution was sought using an analogue computer to model the transfer function of the system; relative acceleration levels between the base and platform were then decreased from 11.5G to 2G with a significant improvement in damping.

A similar experimental arrangement was employed by Hiller *et al* [65] except that a piezoelectric shaker table was used to generate a 7 Hz sinusoidal periodic vibration beneath the base mount. Disturbance signals from the top of the magnetostrictive antivibration module were processed in digital form using a PC and, via a D/A converter, controlled current to the magnetostrictive actuator. With a simple proportional controller, the displacement amplitude was reduced by 6.5 dB after one second of transient response had elapsed. The transmission function was measured, rising from ~50% at low frequencies to ~90% at 90 Hz. This work demonstrated that low-frequency active vibration cancellation can be achieved while maintaining static level control. To provide comparable isolation at 2 Hz, a passive system would require a natural response frequency <1.4 Hz and such a compliant passive system would inevitably lead to difficulties in maintaining static levels. Hiller *et al* fully appreciated that there were inadequacies in the performance of the magnetostrictive device and that sampling times in the control system were restricted to ~200 μs (a sampling frequency of ~4.5 kHz) which affected the system's band width, so improvements in the control system hardware were also seen to be desirable as well as improvements to the control algorithms.

Recently, progress with actuation and control using Terfenol has been reported by Greenough *et al* [50]. Output from two vertically stacked actuators, one being the base generating prescribed vibration, and the other forming an anti-vibration module, has been measured. Initially the responses of each actuator were measured in response to a controlled demand for a triangular waveform. The DC bias to each actuator was changed and from the measured hysteresis in each actuator response, the need for control of the DC bias is demonstrated. The hysteresis from the two actuators was not the same, as expected, because the loading on each actuator was different. The movement of each actuator in response to a specific demand signal was then measured with PID control in operation. However for regulation of movement under variable load and with the properties of Terfenol in mind, a control strategy has been developed which combines the techniques of variable structure and self-adaptive control. The switching properties of the variable structure ensure that a process model is identifiable; the self-tuning property ensures that the variable structure operates along an optimal switching line [51]. The response of the system operated with this control to a simulated stepped vibration input illustrates how the adaptive process contends with actuator hysteresis. To implement the variable, adaptive self-tuning control strategy in real time requires a large amount of computer processing time and a transputer based micro-controller has been developed for this purpose with sampling rates ~10 μs. While in some circumstances PID control may suffice, complex anti-vibration tasks (such as a system with multiple actuators) will inevitably require more sophisticated control. The desired level of control and its achievement can only be judged in relation to the output specifications for a particular task.

8.5.6 Miscellaneous Applications

A prototype magnetometer to measure field strength using Terfenol-D has been described by Chung *et al* [66]. In order to resolve the small changes in length, a He–Ne laser heterodyne interferometer technique was proposed. From the non-linear response to change in cavity length, a resolution of 8.3×10^{-7} μm was expected. A diode laser with PIN detector was used to investigate this technique in practice and examine the dynamic response. The technique was then applied to magnetometry using the reflection from the polished end of a 10 cm long rod of Terfenol. A 1.4 kHz drive field of magnitude 80 A m^{-1} and a DC bias of ~14.6 kA m^{-1} were generated with a solenoid so that 1.28 nm of movement could be detected (corresponding to a d coefficient ~152 nm A^{-1}). The device was capable of measuring minimum field strengths of ~160.5 μA m^{-1} (2.07×10^{-6} Oe). With a portable device in mind, the objective was to keep the overall dimensions of the magnetometer to ~$120 \times 80 \times 25$ mm; even with a 1 mm disc of Terfenol, a resolution of 16×10^{-3} A m^{-1} (0.2 mOe) is achievable.

The concept of a magnetostrictive electric generator has been proposed by Lundgren *et al* [67] in which Terfenol is operated in reverse mode i.e. an oscillating force is applied to the end or rod and the corresponding change in flux due to magnetomechanical coupling induces a current in the surrounding coil. Efficiency is considered in relation to eddy current loss.

Finally in recent papers by Clark *et al* [68] $Tb_{0.6}Dy_{0.7}$ alloys are described which form the basis of a low temperature actuator [69], bringing together several advanced technologies. Instead of using a magnetostrictive rare earth–iron compound at room temperature, Tb–Dy alloys have been fabricated in textured sheets by hot rolling and thermal processing. These grain oriented materials generate magnetostrictive strains ~4000 to 6000 ppm, with a magnetomechanical coupling coefficient $\leqslant 0.7$ at liquid nitrogen temperatures, ideal for harnessing high-temperature superconductors to generate the magnetic fields. Modern techniques can also be employed to generate the low temperatures. This configuration has a clear advantage that the field production is highly efficient and the self-heating problems associated with conventional copper windings are avoided.

8.6 SUMMARY AND CONCLUSIONS

The present work has reviewed the properties of highly magnetostrictive rare earth compounds, provided guidelines for the construction of devices, described analytical techniques to examine material or device performance and reviewed a broad cross-section of applications. Their future success depends on how well the properties can be engineered into working systems at a cost which is commercially attractive, keeping in mind the fact that established technologies may perform equally well. The rolled Tb–Dy alloys mentioned in the last section are the cheapest form of highly magnetostrictive materials but the grain-oriented rare earth–iron compounds are, as yet, the best room temperature materials. Their production costs are not particularly cheap but are expected to decrease as demand increases.

Perhaps one message this present work can convey is that a simple approach to actuation with highly magnetostrictive materials—namely to wind a solenoid around a rod of material and switch 'on'—can be counter-productive. Without carefully considered designs and construction, excessively large powers will be required and mechanical output disappointingly low; this new technology is then all too easily discarded in favour of an inferior alternative which may happen to be in a more advanced state of development. The greatest need at the present time is to build efficient, innovative demonstration prototypes so that the future users can see the potential. Apart from the need for demonstrations, the question remains as to why magnetostrictive materials have taken so long to reach the application stage. There are several possible answers. One is the time it has taken for the materials to be engineered from the discovery of giant magnetostriction in rare earth elements to practical materials for use at room temperature. A great deal of credit is due to Dr A E Clark and co-workers for that successful transition, the leading authors in this field. Another reason is that Terfenol has competitors and it must be seen to have some special advantages before the end user is persuaded to abandon time, effort and investment devoted to other technologies. However, as these advantages are realized in practice, the more acceptable highly magnetostrictive materials will become. Finally there has to be communication between those who already work in the field and potential users who don't even know that Terfenol technology exists. It is hoped that this present work will help to bridge that gap.

The authors apologise for any apparent omissions from their survey. It has been impossible to cover every aspect (one omission, for instance, is a description of power supplies and amplifiers to feed the devices) but it is hoped that no major contribution is missing! The authors would also like to express their appreciation to all those who directly or indirectly made a contribution, particularly to staff in the departments of Applied Physics, Electronic Engineering and Computer Science at Hull University whose shared experiences in this field have been of benefit in the preparation of this article.

REFERENCES

[1] 1946 National Defence Research Council Div. **6** 13 DDC No T1669
[2] Belov K P, Levitin R and Nikitin S 1961 *Sov. Phys.–JETP* **13** 196

[3] Clark A E 1980 *Magnetostrictive Rare Earth–Fe_2 Compounds in Ferromagnetic Materials* ed E P Wohlfarth (Amsterdam: North Holland) ch 7
[4] Verhoeven J D, Gibson E D, McMasters O D and Baker H H 1987 *Met. Trans.* **18A** 223
[5] Davies Jr C M 1977 *US Navy J. Underwater Acoust.* **27**(i) 39
[6] Mason P W 1960 *Piezoelectric Crystals and their Application to Ultrasonics* fifth edn ch 5 (New York: Van Nostrand)
[7] Greenough R D and Schulze M P 1990 *IEEE Trans. Magn.* **26** 2586
[8] Butler J L 1988 Application Manual for the Design of Etrema Terfenol-D Magnetostrictive Transducers (Image Acoustics Inc. North Marshfield, MA) p 28
[9] Kendall D and Piercy A R 1993 *J Appl. Phys.* **73** 6174–6176
[10] Clark A E, Teter J P, Wun Fogle M, Moffett M and Lindberg J J 1990 *J. Appl. Phys.* **67** 5007 (1990)
[11] Moffett M B, Clark A E, Wun Fogle M, Lindberg J F, Teter J P and McLoughlin E A 1991 *J. Acoust. Soc. Am.* **89** 1448 (1991)
[12] Moffett M B, Powers J M and Clark A E 1991 *J. Acoust. Soc. Am.* **90** 1184
[13] Clark A E and Crowder D N 1985 *IEEE Trans. Magn.* **MAG-21** 1945
[14] Callen E and Callen H B 1965 *Phys. Rev. A* **139** 455
[15] Savage H T, Clark A E, Koon N C and Williams C M 1977 *IEEE Trans. Magn.* **MAG-13** 1517
[16] Koon N C and Williams C M 1979 *J. Physique* **C5 40** 194
[17] Sato K, Isikaura Y, Mori K, Clark A E and Callen E 1986 *J. Magn. Magn. Mater.* **54–57** 875
[18] Clark A E, Teter J P and Wun Fogle M 1991 *J. Appl. Phys.* **69** 5771
[19] Prajapati K, Jenner A G, Schulze M P and Greenough R D 1993 *J. Appl. Phys.* **73** 6171–6173
[20] Clark A E, Cheng S F, Teter J P, Wun Fogle M, Huang M Q and Zheng Y 1992 *J. Magn. Magn. Mater.* **104–107**, 1433
[21] Clark A E, Restorff J B, Wun Fogle M and Lindberg J F 1993 *J. Appl. Phys.* **73** 6150–6152
[22] Sandlund L, Fahlander M, Cedell T and Stähl J-E 1992 *Actuator 92, Proc 3rd Int. Conf. on New Actuators* (*VDI*/VDE Tech. Inf.) (*Bremen, 1992*) p 249
[23] Schulze M P 1990 Nondestructive testing and evaluation of magnetostrictive materials, *PhD Thesis* University of Hull, UK
[24] Greenough R D, Jenner A G I, Schulze M P and Wilkinson A J 1991 *J. Magn. Magn. Mater.* **101** 75
[25] Abell I S, Butler D, Greenough R D, Joyce V and Pitman K C 1986 *J. Magn. Magn. Mater.* **62** 6
[26] Meeks S W and Timme R W 1980 *IEEE Trans. Magn.* **MAG-27** 60

[27] Reed I M, Greenough R D and Schulze M P 1993 *IEEE Trans. Magn.* **MAG-29** 2386–2388
[28] McMasters O D, Verhoeven J D and Gibson E D 1986 *J. Magn. Magn. Mater.* **54–57** 849
[29] Clark A E, Verhoeven J D, McMasters O D and Gibson E D 1986 *IEEE Trans. Mag.* **MAG-22** 973
[30] Bi Y J, Hwang A H H and Abell J S 1992 *J. Magn. Magn. Mater.* **104–107** 1471
[31] Greenough R D, Schulze M P, Jenner A G I and Wilkinson A J 1991 *IEEE Trans. Magn.* **MAG-27** 5346
[32] Butterworth S and Smith F D 1931 *Proc. Phys. Soc.* **43** 166
[33] Savage H T, Clark A E and Powers J M 1975 *IEEE Trans. Magn.* **MAG-11** 1355
[34] Earwicker M Private communication
[35] Claeyssen F 1989 Conception et realisation de transducteurs sonar bass frequence à bas d'alliages magnetostrictifs Terres Rares-Fer *PhD Thesis* L'Institut National des Sciences Appliquées de Lyon
[36] Claeyssen F, Boucher D and Faure S 1988 *J. Acoust. Soc. Am.* **83** 520
[37] 1992 Terfenol-D Application notes **5** 3 (Etrema Products Inc., Ames, USA)
[38] Goodfriend M J, Shoop K M and McMasters O D 1992 *Actuator '92, Proc. 3rd Int. Conf. on New Actuators* (*VDI/VDE Tech. Inf.*) (*Bremen, 1992*) p 283
[39] Sewell J and Kuhn P 1988 *Proc. 3rd Int. Conf. on Giant Magnetostrictive and Amorphous Alloys for Actuators and Sensors* ed C Tyren (Marbella, 1988)
[40] Vermet N, Claeyssen F, Wendling P and Grusse G 1992 *Actuator '92, Proc. 3rd Int. Conf. on New Actuators* (*VDI/VDE Tech. Inf.*) (*Bremen, 1992*) p 133
[41] Stähl J E, Adolfsson C, Cedell T and Walter M 1988 *Proc. 2nd Int. Conf. on Giant Magnetostriction and Amorphous Alloys for Actuators and Sensors* ed C Tyren (*Marbella, 1988*)
[42] Claeyssen F, Colombani D and Tessereau A Private communication
[43] Butler J L 1988 Application Manual for the Design of Etrema Terfenol-D Magnetostrictive Transducers (Etrema Products Inc, Ames, USA)
[44] Kvarnsjö L and Svensson L 1988 *J. Appl. Phys.* **63** 3924
[45] Kvarnsjö L and Engdahl G 1991 *IEEE Trans. Magn.* **MAG-27** 5349
[46] Kvarnsjö L, Bergqvist A and Engdahl G 1992 *IEEE Trans. Magn.* **MAG-28** 2623
[47] Tibert H, Bergqvist A, Kvarnsjö L and Engdahl G 1992 *IEEE Trans. Magn.* **MAG-28** 2808

[48] Daley S 1988 PC based self tuning control of an electrohydraulic system *Actuator '88, Proc. 1st Int. Conf. on Tech. Transfer* (*VDI*/VDE Tech. Inf.) (Bremen, 1988) p 185
[49] Raab U and Iserman 1990 Application of digital control techniques for mechanical actuators *Actuator '90, Proc. 2nd Int. Conf. on Tech. Transfer* (*VDI*/VDE Tech. Inf.) (Bremen, 1990) p 22
[50] Greenough R D, Jenner A G I and Wilkinson A J 1993 Actuation and control of giant magnetostrictive materials *Proc. I. Mech. Eng., Machine actuators and control*, Seminars
[51] Parvinmehr A, Wilkinson A J, Greenough R D and Jenner A G I 1989 *Int. Symp. on Adaptive Systems in Control and Signal Processing ACASP89* (Glasgow, 1989) **2** 547
[52] Smith R R and Logan J C 1977 *US Navy J. Underwater Acoust.* **27**(1) 175
[53] Meeks S W and Timme R W 1977 *J. Underwater Acoust.* **62**(5) 1158
[54] Akervold O L, Hutchins D L, Johnson R G and Koepke B G 1977 *J. Underwater Acoust.* **27**(1) 183
[55] Butler J L and Ciosek S J 1980 *J. Acoust. Soc. Am.* **67**(5) 1809
[56] Oswin J R, Edenborough R J, Pitman K C 1988 *Aerospace Dynamics* British Aerospace Technical Journal 249
[57] Asea Brown Boveri (ABB) Seatek, Västeras (Sweden)
[58] Sewell J 1990 Martin Marietta Aerospace (Baltimore USA) Terfenol-D Application Notes (Etrema Products Inc, Ames, USA)
[59] Savage H T, Clark A E and Spano M L 1984 *IEEE Trans. Magn.* **MAG-20** 1449
[60] Roth R C 1992 *Actuator '92, Proc. 3rd Int. Conf. on New Actuators* (*VDI*/VDE Tech. Inf.) (Bremen, 1992) p 138
[61] Vranish J M, Naik D P, Restorff J B and Teter J P 1991 *IEEE Trans. Magn.* **MAG-27** 5355
[62] Dyberg J 1988 *Proc. 1st Int. Conf. on Giant Magnetostrictive Alloys* ed C Tyren (Marbella, 1988)
[63] Adolphson R 1992 *Actuator '92, Proc. 3rd Int. Conf. on New Actuators* (*VDI*/VDE Tech. Inf.) (Bremen, 1992) p 24
[64] Reed R S 1988 Shock isolation using an active magnetostrictive element *Proc. 59th Shock and Vibration Symp.* vol iv (Albuquerque, 1988)
[65] Hiller M W, Bryant M D and Umegati J 1989 *J. Sound and Vibration* **134** 507
[66] Chung R, Weber R and Jiles D C 1992 *IEEE Trans. Magn.* **MAG-87** 5358
[67] Lundgren A, Tiberg H, Kvarnsjö L, Bergqvist A and Engdahl G 1993 A magnetostrictive electric generator *IEEE Trans. Magn.* **MAG-29** 3150–3152

[68] Clark A E, Wun Fogle M, Restorff J B and Lindberg J F 1992 *IEEE Trans. Magn.* **MAG-28** 3156
[69] Clark A E, Wun Fogle M and Restorff J B 1993 *IEEE Trans. Magn.* **MAG-29** 3511–3513

Biographical Details

Tony P Dorey

Professor Tony Dorey has been working in the area of sensors and related electronics since the 1960s following graduating from Cambridge, UK, with a degree in Physics. Most of his work has been concentrated on silicon-based devices and systems for instrumentation and control applications.

Following periods working at the Philips Research Laboratories and Loughborough University during the 1960s, he has been a member of staff at Southampton and Lancaster Universities where he is currently Head of the Department of Engineering. He gained his PhD for early work on the use of MOS transistors as strain sensors which has led him through investigation of the communications and interface requirements of sensors to the design of total systems and their reliability.

At Lancaster University, UK, he has responsibility for a department with teams working in both Electronic and Mechanical Engineering with special interests in the areas of Microelectronics, Mechatronics and Engineering Design. His interest in Actuators as a vital, but relatively little-researched area, stems from his interests in total system design.

John H Moore

1962–5 Imperial College London, Dept of Physics
Graduated 1st Class Honours, Associate of the Royal College of Science
1967–8 University of Birmingham, Dept of Electronic Engineering
MSc Solid State Electronic Engineering

He joined Lucas Group Research Centre as a graduate Research Officer in 1965 and over the next fourteen years worked on many topics including: electric vehicle control systems (converted an Austin Mini Estate to battery drive with electronic controls—the first electric vehicle in Lucas); research into basic ignition processes in IC engines; development of silicon integrated circuits, power transistors and pressure sensors; photovoltaic power systems for remote sites; development of petrol engine management systems and sensors. Staff responsibilities increased throughout this period ending as Manager, Semiconductors and Transducers Department.

In 1979 he moved to Cirencester, Gloucestershire, as co-founder of a new facility within Lucas Diesel Systems focused on the development of electronic fuel controls for passenger car diesel engines. His position as

Manager, Transducer Engineering brought full responsibility for product development and introduction. He held this position until 1987 by which time the Cirencester operation had grown to around 100 engineers and staff. A number of the fuel systems developed during the period are now in high-volume production for leading European vehicle manufacturers.

In 1987 he was offered a secondment to RSRE Malvern as Project Leader of the first Industry/Government National Electronics Research Initiative on the topic of Silicon Hybrids. This initial project, which brought together research staff from ten leading British Companies into a dedicated facility at RSRE, laid the groundwork for follow-on projects in the key technology of Multi-Chip Modules. These circuit construction techniques are now exploited by many of the original participating companies.

In 1989 he returned to Lucas in Cirencester as co-founder of a new Business Unit in the field of Automotive Sensors. Products under development included contactless position sensors for active suspension systems and silicon accelerometers for anti-lock braking controls involving extensive cooperation with both European and US vehicle manufacturers. This Business Unit was eventually incorporated in the Lucas Control Systems Products Division.

In 1992 he moved to Lucas Bryce in Gloucester as Project Manager—Electronic Fuel Systems for large diesel engines. These products, the first of their type, are now in high-volume production for the leading manufacturer of railway locomotives in the United States.

Throughout his career he has maintained a close association with IOP and served on various Institute Committees. He is now a member of Council and Chairman of the Industry Committee. He has served on numerous IOP and IEE Conference Organizing Committees in the fields of Sensors and Automotive Electronics. He is a Member of the Society of Automotive Engineers (USA) and is the author of a number of papers on automotive sensor topics. He holds around 10 patents in automotive electronics and sensors.

David Dawson

Graduated in 1965 in Mechanical Engineering from UMIST, UK, with 1st Class Honours and then pursued an MSc in Machine Tool Technology also at UMIST under Professor F Koenigsberger. Worked for three years in Rolls Royce Diesels, mainly in the manufacturing area, then joined ICI Organics Division for six years, becoming Division Machinery Engineer, responsible for equipment specification and procurement and some specialist design on all new plant. Joined Lancaster University for seven years as northern area Tutor of the Cambridge University Engineering Department's 'Advanced Course in Production Methods and Management'. More recently Lecturer/ Senior Lecturer in Engineering at Lancaster University. Started the first UK 'Mechatronics' MEng and MSc Course at Lancaster in 1988. Continuing interests in engineering design, teaching and consultancy building on extensive experience of industrially linked research with Leyland Trucks, GEC-Alsthom, K Shoes, P&O and smaller companies.

Roger Stanway

Roger Stanway obtained his BSc in Electrical Engineering from the University of Manchester in 1972 and then gained industrial experience as a development engineer with British Leyland (Truck and Bus Division). He returned to the academic world in 1974 and earned MSc and DPhil degrees in Control Engineering from the University of Sussex. From 1981 until 1993 he was Lecturer and then Senior Lecturer in Mechanical Engineering at the University of Liverpool. His main areas of research were the identification and control of vibrations in rotating machinery and engineering applications of electro-rheological fluids. He has published around 70 papers on these topics. In 1993 he moved to the Department of Mechanical and Process Engineering at the University of Sheffield, where he is concerned with dynamics and control teaching and is a member of the Smart Materials Group. He is a Chartered Engineer, and a member of both the IEE and the IMechE.

John L Sproston

John L Sproston received a BSc in Mathematics in 1964 and a PhD in Engineering in 1967, both from the University of Liverpool in the UK. The topic of his PhD studies was the design of aerofoil sections subject to a pre-specified distribution of velocity using the Hodograph Technique in conjunction with numerical methods of conformal transformation. From 1967 until 1969 he was employed by Bristol Siddeley Engines (later Rolls Royce) involved with turbine blade design. Between 1969 and the present time he has been a Lecturer and then a Senior Lecturer in the Department of Mechanical Engineering at the University of Liverpool where he has research interests in the fluid mechanic properties of electrically stressed dielectric liquids and, in particular during the last fourteen years, in fundamental aspects and applications of electro-rheological fluids. He also has an interest in flowmetering of single- and multi-phase fluids. He is author and co-author of more than 70 journal and conference papers and is joint holder of several patents. He has for many years served on the Static Electrification Group Committee of the Institute of Physics in London.

David H Paxman

After graduating from Christs College Cambridge, D Paxman joined the Philips (then Mullard) Research Laboratories in Redhill, England, in 1958 as a physicist working on low-noise amplifiers for use in the first transatlantic

TELSTAR satellite TV transmission experiments. After a year working as a research assistant in the University of Michigan he returned to England to take up work on a variety of subjects including quantum electronic devices (solid state lasers and IR up-converters), solid state microwave oscillators in GaAs and InSb and eventually power devices. The latter work ranged from infrared studies of plasma effects, deep level state examination to device design and fabrication. With the advent of MOS devices this work eventually led to the area of 'Smart Power'. David Paxman is a member of the IEE, and a Fellow of the Institute of Physics. He has served on a number of committees in the IOP, acted as secretary to ESSDERC and at present has been involved with the technical committee of ISPSD"94.

James S McKenzie

James McKenzie received his BSc Honours degree in Computing and Physics from Cardiff University, UK, in 1988. He then obtained an MSc degree in control systems from Sheffield University, UK. From 1990 to 1993 he was involved in research work for his PhD degree at Brunel University, UK, in the area of optical actuation with support from the UK Science and Engineering Research Council and BP. After completing his PhD in early 1993 he worked on a short-term contract for Lucas AEC and now works in The Brunel Centre for Manufacturing Metrology on a Brite/Euram project on optical fibre sensors. His research interests include silicon-based microsensors and actuators and optical-fibre-based sensors.

Kenneth F Hale

Kenneth Hale obtained his BSc Honours degree in Physics from London University in 1956. He then obtained a PhD from London University in 1961 in applications of thin foil electron microscopy. This was followed by a DSc, also from London University, in 1974. He is a Senior Research Fellow in The Brunel Centre for Manufacturing Metrology (BCMM). Before he joined BCMM in 1986 he conducted research for 35 years at the National Physical Laboratory, and its derivative laboratories, the National Maritime Institute, and British Maritime Technology. He was a Senior Principal Scientific Officer at all three organizations. He has published 75 papers on research and development which include 20 papers on optical fibre sensors and is a co-author of a book on dynamic experiments in the electron microscope. He has been a Fellow of the Institute of Physics since 1964.

Barry E Jones

Since 1986, Barry Jones has been Professor of Manufacturing Metrology at Brunel University and Director of the Brunel Centre for Manufacturing

Metrology (BCMM). He is a previous Honorary Editor of *Journal of Physics E: Scientific Instruments* and a past member of the UK Council for National Academic Awards. He was formerly a member of the academic staff at the University of Manchester and then UMIST. His research interests encompass sensors, transducers and actuators, optical and mechanical metrology, automatic inspection, intelligent and distributed instrumentation. He is author and editor of five books and has published over 180 papers and articles. During the period from 1975 to 1993 he has been awarded some 70 research and teaching grants/contracts involving some 35 research staff and 50 PhD/MSc research students. The research work has been undertaken in collaboration with industry and involved many international links particularly in Europe. Professor Jones was awarded his DSc by the University of Manchester in 1985 and holds fellowships of five professional bodies.

Stuart T Smith

Stuart Smith started his engineering career in 1977 as an apprentice maintenance engineer in the rubber industry. Upon completion in 1981 he took his BSc and PhD degrees at the University of Warwick, UK where he lectured until 1994 having attained a senior lectureship. In 1992 he spent a year at the University of North Carolina at Charlotte under the Royal Academy of Engineering Overseas Secondment scheme and returned there to a permanent post in 1994. During his doctoral studies he co-invented a stylus instrument with the late Kevin Lindsey which is now marketed as the Rank Taylor Hobson Nanostep and has since registered seven more patents for instrument designs and published more than 40 papers in this field. He has also co-authored a book with Derek Chetwynd on the Foundations of Ultraprecision Mechanism Design published by Gordon and Breach.

Mehran Mehregany

Mehran Mehregany received his BS in Electrical Engineering from the University of Missouri in 1984 and his SM and PhD in Electrical Engineering from Massachusetts Institute of Technology in 1986 and 1990, respectively.

From 1986 to 1990 he was a consultant to the Robotic Systems Research Department at AT&T Bell Laboratories, where he pioneered research in silicon micromechanical devices. In 1990, he joined the Department of Electrical Engineering and Applied Physics at Case Western Reserve University as an Assistant Professor. He was awarded the Nord Assistant Professorship in 1991, and was promoted to Associate Professor with tenure in July 1994. He is an associate editor of the *Journal of Micromechanics and Microengineering*. His research interests include solid-state transducers, microelectro-mechanical systems, integrated circuit processes, high-temper-ature electronics, thin-film shape memory alloys, modelling of micro-mechanical devices, and mechanical properties of thin films.

Rod D Greenough

After graduating in Physics at the University of Sheffield, UK, he studied the magnetostrictive properties of cobalt-substituted manganese ferrites for a PhD which he was awarded in 1965. After two years as a Research Council postdoctoral research assistant, he moved to take up his present appointment as a member of academic staff at the University of Hull, UK, in 1967. His present research interests remain in the field of magnetoelastic materials but particularly the application of highly magnetostrictive rare earth—iron compounds for devices in practical systems. To date he has made contributions to more than 120 published papers, conferences or seminar presentations and chapters in two books. He is named as co-inventor in six patents.

Ian Reed

With a degree in Applied Physics with Electronics he remained at the University of Hull, UK, to study the characterization of magnetostrictive materials using a plane wave modelling technique. He was awarded a PhD for this work in 1994 and continued to use this technique while employed as a research assistant studying the AC properties of Terfenol-D. This work has been published along with his collaboration in other research areas and presented at conferences and seminars in the UK and Europe.

M P Schulze

M P Schulze is a member of the Institute of Physics (IOP) and a chartered Physicist through the IOP. He received his Diplom-Ingenieur (BS) from the FH-Osnabrück, Germany (1986) in Materials Science and Technology. He continued his studies on the mechanical and physical properties of soft magnetic materials at the University of Hull, UK, where he was awarded an MSc (1988) and a PhD degree (1990). During his three years of postdoctoral work at the University of Hull he developed non-destructive testing methods to evaluate magneto-mechanical materials. He presently works as a Lead Senior Engineer at a materials testing laboratory in Boulder, Colorado, USA. To this date M P Schulze has published 17 scientific papers and co-authored two chapters in books on intermetallic compounds and engineering materials.

Index